AFRICA SOUTH

T0130795

AFRICA
SOUTH

|||||||||||||||||

by Harm J. de Blij

NORTHWESTERN UNIVERSITY PRESS

Open access edition funded by the National Endowment for the
Humanities/Andrew W. Mellon Foundation Humanities Open Book Program.

Harm J. de Blij, Africa South.
Evanston, Ill.: Northwestern University Press, 1962.

ISBN 978-0-8101-3826-1 (paper)
ISBN 978-0-8101-3827-8 (ebook)

Original edition LCCN 62-14295

AFRICA
SOUTH

||||||||||||||||

by Harm J. de Blij

NORTHWESTERN UNIVERSITY PRESS

Open access edition funded by the National Endowment for the
Humanities/Andrew W. Mellon Foundation Humanities Open Book Program.

Harm J. de Blij, Africa South.
Evanston, Ill.: Northwestern University Press, 1962.

ISBN 978-0-8101-3826-1 (paper)
ISBN 978-0-8101-3827-8 (ebook)

Original edition LCCN 62-14295

To Kathleen and Roland

Preface

For more than a decade, Africa has been the focus of world attention as Africans achieved independence. Ghana and Nigeria in West Africa, the former Belgian Congo in Central Africa, Tanganyika in East Africa—one after another these and other former colonies have occupied the center of the stage. The remaining countries and territories not now governed by Africans are concentrated in the last of the regions of the great continent to be affected by the rising tide of political change: the south.

It is not easy to obtain background material on Africa. Gathering information on the racial, social, historical, economic, and political conditions of a continent may involve a laborious search through a variety of works. Many an interested reader following the news from Africa will not have the time and opportunity to do all this reading. The purpose of this book is to provide such basic information on Southern Africa in concise and accessible form. History, population, politics, leaders, and social conditions are discussed, and an effort is made to convey the local atmosphere through descriptions of towns and countryside.

My views are based upon about ten years' residence and study in various parts of the subcontinent. Many of the facts and tendencies described here were observed first-hand, particularly those in education, the press, local politics, and cultural affairs. I spent time as a student and university lecturer in Africa, as an employee of a Nationalist South African newspaper, as a member of the Johannesburg Junior City Council, and as a member of two local orchestras. These ten years do not, of course, constitute a safeguard against bias in fact and error in interpretation. Some readers will disagree with certain of the statements made in this volume, and time may prove their criticism to be sound. I have tried to avoid the emotional aspects of questions on which opinions are expressed and to state as clearly as possible the facts underlying the matter. If there is factual bias in this book, it is not deliberate.

I should like to thank Mrs. Ernestine Todd for her assistance in the typing of the manuscript and Willard Depree, Alan Rowe, Howell Lloyd, Peter Gould, Douglas Gray, and Barbara Simpson for reading various chapters critically and providing administrative help. Their cooperation of course does not imply agreement with the opinions expressed in my book.

I have many friends in Africa, representing virtually all racial groups and all shades of political opinion. Some will be greatly disappointed by the conclusions I have drawn. I hope that they will nevertheless accept my thanks for the time they spent with me in discussions which made this book more accurate than it would have been without their help.

HARM J. DE BLIJ

Contents

Maps

AFRICA SOUTH

The Wind of Change

Ihe political map of Africa changes at a bewildering rate. Since 1946, colonies and protectorates have become independent states, boundaries have been drawn and redrawn, federations have formed and failed. African leaders have risen to power and international prominence, and the "African Bloc" in the United Nations is a force to be reckoned with. No signs appear that the current wave of political activity in Africa is soon to abate. The "Wind of Change," as British Prime Minister Harold Macmillan has called the urge to African independence, will prevail for years to come. Like the trade winds above the equator, the Wind of Change blows southward. It has penetrated Central Africa and threatens to drive a path into the heart of the south.

Africa today, in very general terms, divides politically into three segments: North Africa, Tropical Africa, and Southern Africa. North Africa, with its Mediterranean heritage, is separated from Tropical Africa in more ways than one by the Sahara Desert. There are obvious differences between Morocco, Tunisia, Lybia, and Egypt, on the one hand, and Ghana, Nigeria, and Liberia, on the other. The French-Algerian conflict appears much less "African," for instance, than events in former French Guinea, in the former Belgian Congo, or in Kenya. Although it cannot be denied that Mediterranean Africa exerts a religious influence in West Africa, the essential differences on either side of the Sahara remain.

Tropical Africa includes the majority of those African states which in recent years have attained their independence. Here, also, lie several of the countries which have long been independent, notably Ethiopia in the "Horn" of the east and Liberia on the west coast. The countries of the French Community, such as Niger, Chad, and Gabon, are part of this region, and the products of British colonial policy, Ghana and Nigeria, vie for dominance in the vastness of the west. The two southernmost countries falling into this general category of independent states in Tropical Africa are the Congo and Tanganyika, which attained their new status in 1960 and 1961, respectively.

Below the Congo and Tanganyika lies Southern Africa. This region includes the Federation of Rhodesia and Nyasaland, as well as Angola, Moçambique, and the huge area subject to the hegemony of the Republic of South Africa. Southern Africa is the last major stronghold of European colonialism, where white minorities continue to rule African majorities. The Africa of the past and the Africa of the future here come face to face. The Congo and Tanganyika afford a haven for rebels and refugees from Southern Africa.

The Wind of Change does not always manifest itself in the same manner. Peaceful political bartering goes on in some areas. Strikes and riots occur in others. Elsewhere, violence and strife grow worse, diminishing the chances for interracial cooperation. The more Southern Africa resists change, the more severe the consequences for the future.

| The Buffer Zone

A map of Africa below the equator shows that of all the countries and territories in Southern Africa, only three actually border the independent African states to the north. Angola, the Federation of Rhodesia and Nyasaland,

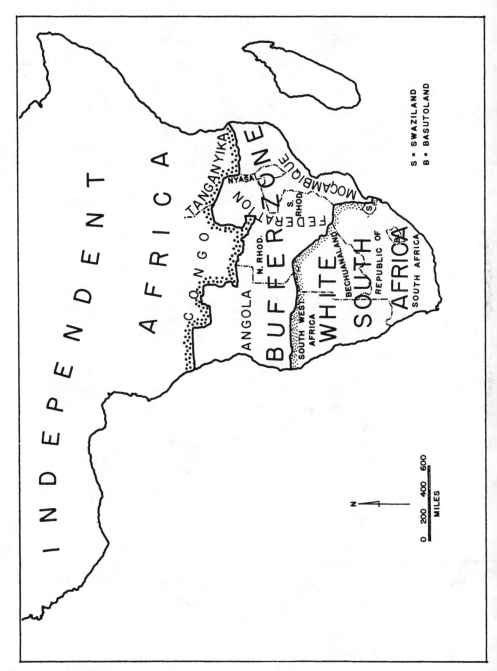

MAP 1   *Subsaharan Africa*

and Moçambique effectively separate the Republic of South Africa and its immediate neighbors from the current sources of the Wind of Change. Thus the two extremes in Africa have not yet come face to face across a political boundary. While Congo and Tanganyika represent one extreme in the form of African nationalism and complete independence, South Africa is the other extreme, in that white nationalists rule a country in which the African majority has no voice. The three northern countries of Southern Africa thus perform the function of a buffer zone. It is the avowed aim of all Africans everywhere, virtually without exception, to help bring about the political emancipation of the non-whites of the Republic of South Africa. It is the intention of the white Nationalist government in South Africa to keep these non-whites segregated from the white population and to maintain as many forms of control over these people as possible. There are many millions of Africans in the now independent African states who would act to interfere in South African affairs. There are many white South Africans who would defend their position to the end. The South African armed forces are the best equipped and most powerful in Africa, and any such rebellion as has been occurring in northern Angola would have disastrous consequences for the rebels and perhaps even for the country which they used as a base of operations. The existence and survival of this buffer zone in Southern Africa may, therefore, be a crucial matter in the maintenance of peace as the political evolution of the south proceeds.

What might be called the core of the buffer zone is the Federation of Rhodesia and Nyasaland, made up of the three territories of Southern Rhodesia, Northern Rhodesia, and Nyasaland. This Federation was established on September 4, 1953, against the will of the vast majority of Africans in each of the three countries, and it has survived political turbulence since that date. There are considerable internal contrasts in the Federal State. Southern Rhodesia is separated from the two other territories by the

Zambezi River, which, as will be seen subsequently, has always played a dividing role in Central Africa. Southern Rhodesia, with its mineral wealth and agricultural potential, attracted a large number of white settlers, and prior to federation its policies closely resembled that of its neighbor South Africa. Even today, though there is some interracial cooperation, many white Southern Rhodesians militantly favor the absolute retention of all control over this country by the white minority. On the other hand, Northern Rhodesia and Nyasaland attracted a smaller white population, and until federation took place, these two countries seemed set upon the road toward self-government. It is of course in Northern Rhodesia and Nyasaland where African agitation for independence has been strongest, and several African leaders, such as Hastings Banda, have gathered support among the Africans in their fight against federation. The Wind of Change blows steadily here, but it has not yet succeeded in the destruction of the "white man's Federation"—the immediate goal of Africans in these parts.

Whatever the future of the Federation, its present character as a buffer is unmistakable. In the north, there is considerable scope for African political activity, and the contrast between conditions in Tanganyika and Nyasaland is not such as to resemble that between Tanganyika and South Africa. Hence, there is a form of transition without excessive abruptness between the independent states and the northern components of the Federation. In the south, the contrasts between Southern Rhodesia and South Africa are likewise mild. In Southern Rhodesia the African has achieved more politically, but not *much* more, than he has in South Africa.

The great contrast is within the Federation itself, divided sharply at the Zambezi River. There are historical reasons for this. Southern Rhodesia, like South Africa, was conquered by the white men who then settled in large numbers. North of the Zambezi, the region was peacefully penetrated for the most part by missionaries, traders, and

British officials who set about terminating the slave trade and maintaining, as much as possible, the African tribal arrangements. In the true sense of the word, Nyasaland and Northern Rhodesia were "protectorates." This initial contrast between north and south in the present Federation, and the subsequent policies of land alienation in the south and preservation of land for the Africans in the north, sowed the seeds of division. Since the early 1950's, this diversity has been combined in a country sometimes referred to as the "Disunited States."

The central buffer zone is flanked by two Portuguese territories, Angola in the west and Moçambique in the east. Portugal for centuries failed to develop the economy of these overseas possessions to any significant degree and totally neglected education, health, and any idea of general progress there. Although some change may now be noticed in Portugal's attitude toward these two huge areas, the effects of hundreds of years of backwardness are not eradicated in months. Angola and Moçambique may be referred to as the weak flanks of the buffer zone, for the very policies which were designed to keep these areas isolated and easily controlled now make them vulnerable. They maintain their function as buffer territories largely because of this unfortunate legacy, and even today it is not certain to what extent they have been affected by the changes occurring everywhere. Few reasonably accurate news reports emanate from these areas. In the eyes of the world, given only censored news by the Portuguese, they seem to have remained unchanged to an even greater degree than they really have. The Portuguese insist that the long-time lack of reports of racial friction in Moçambique and Angola is simply a reflection of harmonious conditions. They failed to conceal the terrors of the 1961 uprising in northern Angola, but the real conditions which gave rise to this and other desperate rebellions are still not well known.

Moçambique and Angola are both territories of great

contrast, with a special character bearing on their function as sections of the buffer zone. Until August, 1961, when all was changed by a decree from Lisbon, Africans in Angola and Moçambique were subject to a system of forced labor only one small step removed from slavery—a condition unheard of anywhere else in contemporary Africa. On the other hand, if an African managed to fulfill a set of requirements, he could elevate himself to virtually full citizenship and, in fact, could count on being accepted virtually as an equal by the local white Portuguese citizenry. Details of the labor-conscription system have always been hard to come by, but the Portuguese point with pride to the equality enjoyed by the African who has become a full citizen. Indeed, it is not likely that an African, however well educated and wealthy, would enjoy such privileges in, for instance, the neighboring Federation, even though labor practices common in Portuguese Africa were unheard of in the Rhodesias and Nyasaland. The Portuguese, by exploiting the propaganda advantages of their system, managed for a long time to secure the sympathies of even those who opposed colonialism elsewhere. Among Africans in Africa itself, the true conditions in the two territories became known relatively recently.

How ironic it is that the few Africans who managed to achieve equality under the Portuguese thereby became citizens of a dictatorship. Portugal's government, even in its European homeland, is not known for patience with disloyal citizens. The army and navy act as especially well-armed police, and events in recent elections have shown that there is not much scope for an opposition party. More so in Angola and Moçambique, few dare express (other than privately) dissatisfaction with the Lisbon regime. These are the conditions which pertain to the whites, the full citizens of Portugal. It is, therefore, not surprising that no African political organization, hinting at African nationalism, has been permitted to arise. Portuguese officials will state that the Africans really believe themselves to be Portuguese and that African nationalism finds no fertile ground in Overseas Portugal. When the police find an African who is so impolite as to contradict this, he disappears from the scene. Thus, in the same country where an African has an avenue to achieve virtual equality, he is unable to express his political beliefs should they be contrary to the state doctrine. While the first factor long protected Portugal against criticism of her colonial policies, the second has delayed a process which elsewhere in Africa helped lead to political emancipation: the evolution of African political organizations in which Africans could gain experience in leadership and which in this case would have permitted the African majority to barter with the Portuguese from a point of strength. A consideration of the political situation in all political units of subsaharan Africa will show that the conditions prevailing in Angola and Moçambique are nowhere paralleled except in South Africa where, however, African and Indian nationalist organizations were banned only recently.

The suppression of political organizations in Moçambique and Angola may for the present aid the Portuguese. On the other hand, it eliminates one avenue, and a very desirable one, along which the Wind of Change might have penetrated. The Africans know that they are jailed or banned for political agitation and that they will be shot if they riot or strike. The only course of action which remains for them is that of armed rebellion such as broke out in northern Angola in March, 1961. For this, equipment and supplies are necessary, and, if the revolt is to be successful, a great deal of organization. It would appear that African guerillas will not for some time be able to withstand a European army. As long as this remains true, Angola and Moçambique are likely to retain their place in the buffer zone. If, however, that army should be required to maintain order in the homeland, the situation would change rapidly. Under government by dictatorship, this development could occur at any time. This is a major factor in the vulnerability of Moçambique and Angola.

| *The South*

As the Federation of Rhodesia and Nyasaland dominates the buffer zone, so the Republic of South Africa is the major power of the south, being at the same time the buffer zone's main beneficiary. South Africa is white man's Africa, where just over three million whites (more than in all the rest of subsaharan Africa) tightly control every aspect of politics and economy. South Africa is famous today for its gold and its policies of absolute racial segregation. Most remote from the effects of the Wind of Change, South Africa is the most populous country in Southern Africa, and economically the most advanced. It virtually controls the fortunes of the other territories south of the buffer zone. Since 1921, it has maintained a mandate over South West Africa, and it has made no steps to relinquish its control over this area. Other mandate territories and trusteeship areas in Africa have become self-governing, but South West Africa is treated almost as a fifth province of the Republic. South Africa, clearly against the will of the United Nations, has applied its principles of racial segregation in South West Africa and has given the white occupants of the territory representation in the Republic's House and Senate while withholding any such form of representation from the Africans. By virtue of its control over South West Africa, the Republic's power now extends from the Cape to the Zambezi, driving South Africa's sphere of influence far to the north of the borders of the Republic proper.

Three other territories in Southern Africa are likely to come to world attention in the future. These are the so-called High Commission Territories of Basutoland, Bechuanaland, and Swaziland. They lie immediately adjacent to South Africa, and the Republic has always exerted a considerable influence in each. Bechuanaland is almost enclosed by South Africa and South West Africa. Swaziland lies in the southeastern corner of the Transvaal province of South Africa and is also virtually surrounded by Republic territory, possessing only a short border with the Portuguese area, Moçambique. Basutoland lies entirely within South Africa, and has no borders other than with the Republic.

Each of these three territories was once a British protectorate, and even today they are still occasionally referred to as the South African protectorates. Actually, Britain has been attempting to guide these countries toward a greater degree of self-government without thereby annoying South Africa to the point of interference. In the process, Basutoland has risen to the stature of colony, and important changes have recently taken place also in the two other areas. South Africa has long cast covetous eyes upon these three High Commission Territories and has repeatedly demanded that their government be transferred to South African hands. Although they do not constitute a threat to South African security, the development, on the Republic's very borders, of African independent states is not appreciated and may not even be tolerated. Here is another potential source of friction.

The two giants in southern Africa are the Republic of South Africa and the Federation of Rhodesia and Nyasaland. In terms of total population, economic progress, and many other factors they far exceed all the other countries in importance. Both, however, lack cohesion. The diversity within the Federation is perhaps not quite matched in intensity by the internal disunity of the South African Republic, but in both cases the very existence of the state has been threatened. In the Federation the Africans of the north are determined to break away, and in South Africa the province of Natal has voiced intentions of secession in several instances. Meanwhile, many forces exert themselves throughout Southern Africa. Portugal of course regulates in Angola and Moçambique. Britain influences the Federation, the High Commission Territories, and even to some extent Natal. India's government has shown concern for the fate of the people of Indian descent in South Africa. Appeals go from Africans in South West

Africa and elsewhere to the United Nations. African nationalists attempt to break up the white man's political
organization. Afrikaners (white South African nationalists) generate hostility by separating themselves not only
from Africans, Asiatics, and "Colored" (persons of mixed
blood) but also from English-speaking whites in their
country.

| Fact and Friction

While it is useful and often necessary to treat Southern
Africa as a single geographical and to some extent political,
economic, and cultural unit, one must guard against over-
simplification. Angola and Moçambique, South Africa and
Southern Rhodesia, Basutoland and Swaziland—they each
have their own physical, historical, economic, and socio-
political characteristics. Boundaries of long standing give
the territories and countries of Southern Africa a feeling of
individualism.

One must guard also against reading too simple a moral
into the history of Southern Africa and expecting a precise
kind of change to take place inevitably, uniformly, and simultaneously. In Africa today medieval colonialism survives
adjacent to African nationalism. Exploitation, detribaliza-
tion, urbanization, and emancipation have unpredictable
effects. White supremacy and integration prevail side by
side. Black nationalism and white nationalism, with their
respective brands of irrationality, mingle.

Understandably, the news which the world receives from
Southern Africa emphasizes friction and conflict. Fact and
friction go hand in hand in this area. To get a full sense of
the life lived in Southern Africa, one must know the factual
background.

FURTHER READING

ADAM, T. R. *Government and Politics in Africa South of the Sahara.* Random House: New York, 1959.

ALEXANDER, L. M. *World Political Patterns.* Rand McNally: Chicago, 1957.

AMERICAN ASSEMBLY. *The United States and Africa.* Columbia University Press: New York, 1958.

EAST, W. G., AND MOODIE, A. E. *The Changing World.* World Book Company: New York, 1956.

GOULD, P. R. (ED.). *Africa—Continent of Change.* Wadsworth: Belmont, Calif., 1961.

GUNTHER, J. *Inside Africa.* Harper: New York, 1955.

WALBANK, T. W. *Contemporary Africa: Continent in Transition.* Van Nostrand: Princeton, 1956.

The African Who's Who Diplomatic Press, London, 1960.

The Impact of Europe

II

Bushmen and Hottentots

The present population of Southern Africa is composed of the representatives of several racial groups. Chronologically, the Bushmen and Hottentot peoples preceded all the others. Known as the Khoisan peoples of Southern Africa, the Bushmen first and the Hottentots subsequently occupied most of the subcontinent. Today, neither people survives in great numbers. There are perhaps 10,000 remaining Bushmen, scattered through the drier parts of the region. Perhaps many centuries ago they numbered in the hundreds of thousands, but even before the first white men set eyes upon Southern Africa, the Bushmen were being driven into the most remote and inhospitable parts by their immediate successors on the local stage, the Hottentots.

Although concentrated today in the Kalahari Desert, the Bushmen have left indisputable evidence of their former dispersion throughout the south of Africa. All over the subcontinent, from the Cape to Southern Rhodesia and from Swaziland to South West Africa, in caves and on protected rock faces, there are "Bushman paintings." Although the Bushmen no longer practice this art, they still do prepare the coloring matter they employed and use it for other purposes. Bushman paintings in many cases are of great beauty. They depict the mode of dress, hunting scenes, and other aspects of Bushman life, and there could hardly have been a better record left by people who did not write. Bushman paintings, which are still being discovered throughout Southern Africa, have caused much excitement. In South West Africa in 1918, at the Brandberg (called Forsaken Mountain by the local Nama people), a very remote part of the country, is a sheer granite wall upon which the Bushmen painted, and among the paintings is that of a white woman. The painting is so clear as to permit inspection of the facial characteristics, which seem decidedly non-African. Did the Bushmen come into contact with whites who were there but left no trace? Reports have recently come from Basutoland of Bushman paintings in which the figures wear Egyptian dress. Contact with the Arabs has been suggested as a possible explanation of the "white woman of Brandberg." Another explanation, though less exciting, comes perhaps closer to the truth. Bushmen, like many other African peoples, paint their bodies at certain times. It is entirely possible that the portrait at the Brandberg is one of a Bushman woman thus decorated. However, it must be added that while the animals in the Bushman paintings are in many cases excellent images of the real subjects, the people do not resemble the Bushmen known to us today at all. Much is yet to be learned about the remarkable art of the early Bushmen.

The present-day Bushmen are as interesting as the paintings of their forefathers. They are small in stature and have been described as the Pygmies of the desert. They live the life of nomadic hunters and collectors but have never learned to store food for time of want. Thus they suffer malnutrition and terrible hardship during the droughts, while leading a somewhat easier existence during the wet season. They protect themselves by smearing their bodies with a layer of animal fat, which collects dust and hardens; washing is unknown. Their skin color is a light yellowish-brown, and their dark hair is extremely curly. The skin itself wrinkles intensely, especially in the face.

Although they live in a country where the climate is harsh, the Bushmen do not build permanent shelters. The only protection of this kind they have is a crude assemblage of branches and perhaps some skins, which acts as a windbreak and in the lee of which they sleep.

The main problem facing the surviving Bushmen is the perpetual shortage of water in the areas in which they live. Most of the remaining Bushmen live in Bechuanaland, a country which lies almost entirely within the Kalahari Basin, one of the driest places in all Africa. The Bushmen must depend upon a relatively limited number of water holes found where depressions in the land bring the water table up to, or close to, the surface. Man and animal rely upon these water holes for survival, and since man in addition depends upon the animals for food, Bushmen never set up permanent shelters near water holes, for fear of frightening the animals away. It is here at the water hole that the Bushman often stalks his prey or perhaps poisons it by temporarily poisoning the water supply. Usually, Bushman shelters are found some distance away, radiating in all directions from a water hole, and water holes may be shared by many families.

Bushmen live in groups of some dozens of individuals, commonly a number of families together. Large agglomerations of people are prevented from forming by the scarcity of food and water. Agriculture has never been a Bushman occupation, and although Livingstone is reputed to have taught irrigation to Bushmen he met near the Zambezi River, the practice did not spread. The Bushmen are excellent trackers and hunt with bow and arrow, spear, poisoned spear, and knobkerrie, a stick with a heavy knob at one end. Their legendary accuracy with the spear has not always been verified by demonstration, but the Bushman hunter's determination and tireless persistence are incredible. He will stalk, wound, and then track an animal for days, if necessary, wandering many miles from shelter and clan. Bushmen are also known to stalk animals by imitating their appearance. With the aid of ostrich feathers, for instance, they contrive to resemble an ostrich and then quietly approach the unsuspecting animal.

Although a fascinating people, the Bushmen are no longer significant numerically in Southern Africa. They have been succeeded by invading peoples and have been decimated by each. The first of these were the Hottentots. Precisely when the Hottentots began competing with the Bushmen for the domination of the south is uncertain, but the latter were being driven out of the fertile Cape even before the white man first reached the subcontinent. The Hottentot's language, beliefs, and legends are related to those of the Bushmen. Although the Hottentot is usually somewhat taller, he resembles the Bushman in general appearance. The Hottentots, however, may be said to have been more advanced than the Bushmen at the time of their conflict, for they were a pastoral people with a more complicated tribal organization. In addition, they possessed huts of superior construction which were dismantled and reassembled as the people followed their animals to new pastures. Nevertheless, they shared with the Bushmen their nomadic habits, and neither people practiced agriculture. The Hottentots had a working knowledge of iron and copper, but they were using mainly stone implements at the time of the arrival of the first whites.

The Hottentots, while still in conflict with the Bushmen, faced the vanguard of two new invasions to which they were eventually to succumb. From the north, the Bantu peoples were spreading across Central Africa and into Southern Africa, and from Europe, the white man's ships brought settlers to the coasts. Although the Bantu did not reach the present Cape Province of South Africa until after the arrival of the white man, Bantu peoples occupied other parts of Southern Africa long before the first Portuguese ship sailed past the southern tip of the continent. The Bantu wave of migration was massive, and today the Bantu are the most numerous people in the subcontinent. They spread into the Transvaal, Swaziland, Natal, and eventually into the eastern Cape, while also penetrating into northern South West Africa. The Bantu immigration was not that of a single, homogeneous group. Although they have certain somatic characteristics in common, the Bantu are subdivided into a large number of peoples, and those who were invading South West Africa, for instance, may have been very different in terms of social organization and mode of life from those who were taking a more easterly route to the south. Some differences are obvious from the amount of conflict that occurred, not only between Bantu and Hottentot, but also between various Bantu peoples. Those who reached the Cape first were the Xosa people, a semi-sedentary pastoral people who, like the Hottentots, kept cattle. In what is today Natal, the Bantu people were united, soon after their arrival, into the Zulu nation. Under strong leadership, these Zulu were for decades to rule much of Southern Africa by terror. In northern South West Africa, the Herero became a powerful people, who waged war with the Hottentots for several decades. These—the Xosa, Zulu, Herero, and many others —are all Bantu peoples. Their division is as strong as ever, but they are united by a common purpose. These Bantu are the Africans now striving toward political emancipation, after having for centuries been dominated by the white invaders from Europe. About these Bantu Africans, much more will be said subsequently.

| The Portuguese Land

White men set eyes on Southern Africa before Columbus first saw America. Portugal in the fifteenth century was seeking a way to the Far East, and in 1486 Bartholomeu Diaz landed on the shores of what is today South West Africa. In 1487 Vasco da Gama sailed around the Cape and landed on the Indian Ocean coast of subtropical South Africa. Because it was Christmas at the time of Da Gama's landing, he named the area Natal, a name which it retains to this day. Even though landings were being made, however, there was no major effort at colonization and no great interest in the land and its contents. Southern Africa during the sixteenth century remained an obstacle to be circumvented rather than an inviting prize. Nothing to rival the great struggle for South America and Middle America took place here. Unlike parts of Western and Central Africa, Southern Africa was not a great slave reservoir, and although the slave trade was to affect parts of the south in later years, the remoteness and hostility of the Southern Africans combined to defer these activities here for a long time.

While the Portuguese showed some interest in the coasts of present-day Angola and Moçambique, they continued to avoid the Cape. There were several reasons for this. The Cape of Good Hope is a very stormy coast, and in addition there are treacherous offshore rock ledges upon which many ships have split their hulls. When fair weather appears to prevail, the southwesterly winds may be blowing at gale strength, rendering the whole coast very dangerous. Even today, huge ocean liners each year are prevented by storms from leaving Cape Town. The effects of such conditions upon sailing vessels of only a few hundred tons are not difficult to imagine, and it is not surprising that the Portuguese chose to establish themselves elsewhere.

Beside the storminess and rockiness of much of the Cape's coastline, that were other reasons for the apparent lack of Portuguese interest in the south. The fate which befell Francisco d'Almeida and his party, for example, showed the Portuguese what dangers existed there. D'Almeida was returning from several years' service as a Portuguese official in the Indies, and he chose to anchor in Table Bay, presently Cape Town's harbor. This occurred early in the 16th century, when the Hottentot people dominated this part of the Cape. Apparently, D'Almeida made friendly contact with these Africans, but when he and his men were about to return to their ships, a quarrel broke out and a fight ensued. He thereupon decided to withdraw,

later to return with a larger and better armed party, to
punish the Africans. In this later raid, virtually all D'Almeida's men were killed by the Hottentots. When the few
survivors arrived home, their story spread like wildfire
throughout Portugal. Portuguese sailors consequently
avoided landing at the Cape at all costs.

Neither did the southeast coast hold much attraction.
Its reputation was based largely upon the tragic journey
of Manuel de Sousa. In 1552, De Sousa beached his ship,
because of leakage, on the coast of what is today Pondoland. Aboard was a party of some five hundred men,
women, and children. De Sousa's intention was to make
his way north in the direction of the Portuguese settlements along the east coast of what is today Moçambique.
At that time, the Bantu already occupied parts of the intervening region, which teemed with wildlife. The party
moved northward on foot, the survivors eventually walking about a thousand miles. Only 120 survivors reached
the Limpopo River, wild animals, disease, and skirmishes
with the Africans having taken daily toll. As the party
moved eastward along the Limpopo River, more deaths
occurred, and eventually a mere 20 people, the majority
mentally disturbed, reached Moçambique in May, 1553.
They had spent about one year in traversing the distance.
Like the story of D'Almeida's adventures, the tales of this
disaster reached all corners of Portugal and made a profound impression there. It is small wonder that the Portuguese continued to leave the southern part of the continent
strictly alone.

During this period of Portuguese domination of the
trade routes to the Indies, then, it cannot be said that there
was any significant white immigration into the area here defined as Southern Africa. A few settlements did exist along
the coasts of Moçambique, but these remained small. For
centuries, Portugal confined its interest to the coastal belt
and a few parts of the interior, mainly for purposes of the
slave trade and other crude exploitation. It was nearly 170

years after Diaz' first landing that a beachhead of significant white immigration into Southern Africa was established at Cape Town.

| The Hollanders at the Cape

The 17th century saw the decline of Portuguese power and
the rise of Holland to its "Golden Century." Much of
Holland's wealth was based on its merchant navy and its
possessions in the East Indies, and the Dutch East India
Company was established to coordinate the trade. In 1651,
the Company called for the establishment of a revictualling
station at the Cape, and late that year, Johan van Riebeeck
and a party of men boarded the "Drommedaris" and set
sail for the southern tip of Africa. In traveling to the
Cape, the Dutch ships had to make use of the Canaries
Current and the Northeast Tradewinds; they would then
cross the Atlantic Ocean to the South American east coast,
cross the dangerous Doldrums with the aid of the southward flowing Brazil Current, until, in the general latitude
of the Cape, they would begin to be driven eastward again
in the Westerly wind belt, thus passing close to the tip of
Africa. The tortuous journey from Amsterdam to Cape
Town, if it were a fortunate one, might take as long as
three or four months, and the total distance traveled might
well exceed twice the actual straight-line distance between
the two places. A fresh supply of food and water at the
Cape would greatly alleviate the lot of the sailors and
passengers.

On April 5, 1652, the crew of the "Drommedaris"
sighted land, and on April 6 Van Riebeeck guided his ship
into Table Bay at the foot of magnificent Table Mountain.
It must be remembered that the party was under orders to
set up a temporary revictualling station and that there
were no immediate intentions to set up a Dutch colony
here, although there is evidence that the home government was advised to consider this step. As workers in the
employ of the Company, the party of settlers was to

confine its activities to the setting in motion of the mechanism for revictualling ships; excess production and private
trade with the Hottentots was prohibited. The Hottentots
were the first Africans with whom the new settlers came
into contact, and trouble was not long in coming. With the
settlement hardly 18 months old, Hottentots murdered a
white guard and drove all cattle into the interior, causing
shortages not only for the passing ships but also for the
local party itself. However, Van Riebeeck had been placed
under orders to avoid all conflict with the local inhabitants,
and no reprisals were made. Apparently, relations between
the Africans and the white settlers improved considerably
after the first contact as a result of this policy, and good
relations were generally maintained.

Although the winter rains that year came later than
usual at the Cape, and were unusually violent when they
did come, washing away many of the vegetable gardens
that had been prepared, people soon began to realize the
value of the land they had settled and the opportunities for
private farming and trade. This fact attests to the great
beauty and richness of the Cape, for death and disease were
rampant during the early stages of settlement, a period
from which the small party was a long time in recovering.
As early as April 28, 1655, Van Riebeeck wrote home to
the governing Council of Seventeen that the settlers would
"in due course break altogether with Holland and one day
make this place their fatherland." In 1657, indeed, a small
number of people resigned from the employment of the
Company and settled privately along the slopes of Table
Mountain. However, they were not allowed to sell products to any agent other than the Company, and then only at
fixed Company prices. These limitations proved unaccept-
able to these first real colonizers, and they appealed to Van
Riebeeck to lift them: "We wish to be no Company's
slaves."

Labor became a problem as the settlement expanded.
"You might as well look for jewels in a hogsty as artisans
among this barbarous generation," wrote a disgruntled

settler. The first slaves were imported as early as 1658, and thus a new element was added to the heterogeneity of the Cape. The slaves came from both coasts of Africa, west and east, while others came from the islands of the Far East. Some came from Madagascar. While the import of slaves alleviated one problem, the expansion of settlement created another. The Hottentots' pastures were being alienated by the expansion of the white farms. Van Riebeeck attempted at first to prevent aggravation of these difficulties by establishing a boundary beyond which such white settlement was not permitted to expand. This, of course, proved impractical, and the first armed conflict broke out between whites and Africans. As in countless subsequent instances, the cause of friction was land.

When Van Riebeeck left the Cape for the Indies in May, 1662, he had laid the groundwork for permanent white settlement at the Cape. Apart from developing an economy, he had also established fortifications, achieved some degree of political advancement for the farmer-settlers, treated the Hottentots with a degree of fairness under the circumstances, fought a successful war against them while insisting that efforts to Christianize them continue, and organized the importation of slaves. Many of his works have some effect today. The slaves have played a significant role in establishing the racial composition of the South African population. Cape Town remains the gateway to Southern Africa. Even the fortifications built by Van Riebeeck are in use today by the South African Defense Department.

Van Riebeeck began what was to be some 150 years of Dutch domination of the Cape. During this century and a half, the Cape expanded, though not greatly compared to the expansion of white settlement in such areas as Australia and North America. However limited its expansion, the Cape took on a very definite character. Van Riebeeck had arrived with orders to establish the form of government already in existence in other areas possessed by the Dutch East India Company. Thus, in the beginning, there was a Commander (later Governor) who, appointed by the Company, was responsible for the entire settlement. The Commander was assisted by a Political Council comprised of the most important people at the Cape. In this Council Van Riebeeck included two of the settler-farmers who had left the employment of the Company and wished to farm permanently and independently at the Cape. Soon after the departure of Van Riebeeck, the Dutch began to recognize the importance of the Cape as a protector of the overseas trade. Emigration to the Cape was encouraged, and a large population there was now seen as an advantage. With the growth of the Cape in population and occupied area, government became a more complex matter. In 1679, during a period of relatively rapid expansion, Simon van der Stel was appointed Commander at the Cape. It was during Van der Stel's government that French Huguenot refugees began to arrive at the Cape: in 1688 a party of 200 came, and in 1690 the total white population of the settlement was some 800. The Commander made a policy of scattering these French refugees all over the settlement and prohibiting the use of the French language, but they left a distinctive mark on the colony nonetheless. French contributions to the South African scene include the world-famous Cape vineyards and wines, a nasal twang to the Afrikaans language, and numerous family names in South Africa, some among the most common of all, such as Joubert, Celliers, Marais, Du Toit.

Sporadic border troubles erupted during Simon van der Stel's rule at the Cape, but they were not severe, and they became an accepted consequence of expanding settlement. No contact had been made with the Bantu, and the Hottentots were aware of the reprisals that could occur after excessive resistance to land alienation. Hence the burghers of the Cape were mainly concerned with matters other than friction with the Africans. Simon van der Stel's rule ended in 1699, and his son, Willem Adriaan, succeeded him.

With Willem Adriaan van der Stel began what may be called the dark decades of the history of white settlement in Southern Africa. Though a capable colonizer and excellent farmer, the younger Van der Stel proved corrupt and unreliable, and he pursued an unscrupulous campaign of personal gain at the expense of the farmers. In addition, his activities coincided with a reversal of policy on the part of the Company, which wished to reduce the costs of the Colony, which, it considered, was becoming an excessive burden. Faced with hardship, some farmers took to a form of nomadic cattle-herding and trekked into the interior, mainly along the river valleys between the Cape Ranges. Between 1700 and 1730, these people pushed the frontier farther than ever before, they separated themselves from the "civilization" of the Cape, and they removed themselves beyond the limits of effective government. To the north and east they moved, in small numbers but possessed of an individualism that appears to reassert itself at times in the modern Afrikaner.

While the Council continued to sit in Cape Town, the nearby town of Stellenbosch became the center of administration for the areas farther into the hinterland where the free burghers resided. Stellenbosch became the center for the farmers; Cape Town began to attain the characteristics of an urban center, and the population there had entirely different interests. The self-sufficiency of the Cape, the lack of economic stimulus, and the control of the waning Company combined to retard development. In 1781, the French navy occupied Cape Town for a period of four years, providing some stimulus to the place. This was not, however, because France saw the Cape as a desirable colony to possess, but to keep the British from occupying it. With the threat over, the French left in 1784.

The events taking place in the interior during the last half of the 18th century are of some importance, for the trekkers who had herded their cattle into these remote parts were coming into contact not only with marauding Bushmen in the north but with Bantu in the east, and a series of conflicts ensued. For some considerable time, the Bushmen had much success in harassing the white farmers, but by the end of the 18th century they were beginning to lose the unequal battle against the well-armed groups of cattlemen. In the east, a series of conflicts with the Bantu occurred, known as the Nine Kaffir Wars. Sporadic contact had been made with the Africans of the east since the early part of the 18th century, but the Company attempted to prevent serious conflict by prohibiting trade between black and white. The African people in question were the Xosa, and, like the whites, they were a cattle people in search of good pastures. While the white cattle-raisers were expanding their holdings toward the east, the Xosa were on their way south and, eventually, west, along the same well-watered slopes of the Cape Ranges. The effectively governed territory of the Company did not extend nearly as far as these areas, and the laws established to prevent friction between the races went unheeded. The major contact took place along the Great Fish River, and efforts were made to name this river the boundary between the conflicting groups and their animals. The Great Fish River runs south-southeast from the Cape Ranges to the Indian Ocean and reaches the sea approximately at the point where the South African coast turns northeast. The river lies in a fertile if limited basin, and it is not a matter for surprise that both races coveted the territory. Cattle raids were common on both sides. In 1778 Governor van Plettenberg attempted to obtain agreements from the African chiefs that the Great Fish River should become the permanent and final boundary and that no crossings should be made by either group for any reason. However, he had only limited success in getting local chiefs to agree to this suggestion, and a major conflict broke out in 1779. Thousands of cattle changed hands in this First Kaffir War, and many people as well as animals were killed.

Much of the course of conflict between black and white in the outlying districts of the Cape can be blamed on the failure of the Company to administer the outlying territories or to make an effort to keep them effectively controlled. Outside Cape Town, very few "districts" had been proclaimed: Stellenbosch was first and has been mentioned. Only in 1746 was the District of Swellendam established, and the District of Graaff Reinet was not delimited till after the First Kaffir War. Subsequently, Tulbagh was proclaimed to the north of Cape Town, including those areas where most trouble had been encountered with the Bushmen. The frontiersmen felt that the failing Company had done nothing to protect them, and the Districts became the seats of discontent and political agitation. Company officials were defied and even expelled. There was no great dismay, even at Cape Town itself, when the French took possession of the settlement in 1781. Holland's decline and internal division were reflected at the Cape, and after the invasion of Holland by the French in 1794 and the exile to Britain of the Dutch government, the days of Dutch domination of the Cape were numbered. At the request of the expelled Dutch government, Britain established a caretaker government at the Cape in 1795, and so began the eight years of the first British occupation of the settlement.

Like the French before them, the British during this first occupation were by no means unpopular with the local residents. For the first time, the local government was an effective government, and the Cape settlement was ruled from Cape Town rather than from Europe. Restrictions on market prices were lifted, and Cape Town became a free city. While the western Cape benefited, however, the interior did not change from the better, and with the settlers there the new government was as unpopular as the old. Two more wars with the Africans took place, and as always they had pastures and cattle as their cause. Efforts by officials of the caretaker government to end the conflict were only partially successful, although things were fairly quiet at the end of the first British occupation in 1803.

In 1803, the British handed the government of the Cape back to the Dutch, and the new Dutch government sent officials who were much aware of the changes in European

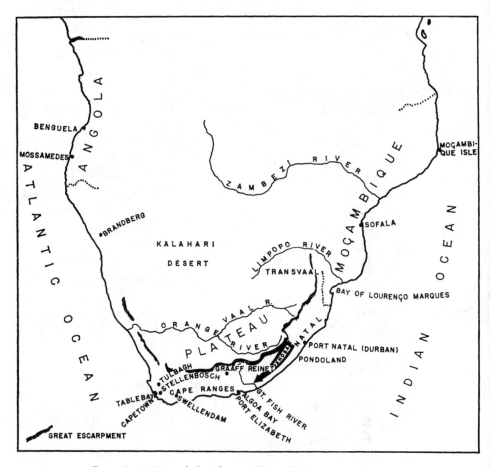

MAP 2 *Location Map of Southern Africa Prior to the Development of the Modern Political Framework.*

thinking at the turn of the century, changes which had left the Cape untouched. The new administration, like the old, was disliked. It did not, though, have much time to implement its plans, for hostilities broke out in Europe once again, and this time Holland was allied with France. The British in 1806 took the Cape once again with a strong fleet, and this time occupation was permanent.

Thus ended Dutch mastery at the southern tip of Africa. From the point of view of the colonizers, Dutch rule at the Cape had not been the success it was in the East Indies. In reality, two Cape settlements had been created with little in common; the frontier and the town were poles apart long before the British arrived to make these differences even more strongly felt. Overseas rule had not been successful, and rebellion had been stimulated. The interior was never ruled effectively, even though opportunities for massive expansion had been thwarted by the policies of the Company. By 1800 the boundaries of the District of Graaff Reinet barely reached the Orange River, and the entire Cape settlement, including its most far-flung frontiers, covered an area smaller than the present Cape Province of South Africa. The Hottentot people had been subverted to slavery and were decimated by diseases brought from Europe; their more or less elaborate social organization was completely destroyed, and there was a good deal of intermarriage. By contrast, the more primitive Bushmen proved more tenacious and fought violent battles with the advancing frontiersmen. They were never subverted to slavery and chose death over capture and imprisonment. All the same, they were driven farther into the Kalahari Desert. Meanwhile, the scene of battle shifted to the east, where Bantu and white man were disputing the cattle lands of the southeastern Cape.

Nothing attempted at the Cape had really met with complete success. There was a good deal of internal division in the settlement, and the total white population at the time the British took over has been estimated at only 25,000. The political status of the colony as a whole and the political status of the inhabitants were frequently uncertain and always unsatisfactory to many. Economic restrictions had prevented what might have been rapid progress in many spheres of activity, and the brief stimulus afforded by the foreign occupations suggests what might have been achieved without these limitations. The important feature of white settlement at the Cape during the first

150 years lay in the particular brand of individualism developed in the farmer and cattleman of the outer districts. This individualism was soon to assert itself in a manner which was vitally to affect the entire course of history in the subcontinent and which on occasion still plays a role today.

| Thirty Years of British Rule

Severe problems faced the new British administration, problems created not only by potential racial conflict along the borders of the newly acquired settlement but also by internal division, a conflict of ideas and interests, and a certain bitterness and resistance on the part of many settlers toward any unsympathetic government, whether it be Dutch or British. These problems would have tested a traditional government. However, the people who were to rule the Cape, like their Dutch predecessors in the final years of Dutch rule, had been influenced by the changes that took place in Europe at the end of the 18th century, and they strove against inertia to use new ideas. In addition, British government inevitably brought with it the imposition of alien law and language, and the first 30 years of the new administration were filled with strife.

One of the early sources of friction was the treatment of the Africans in the Colony. In Britain, the movement against slavery was gaining momentum, and the missionaries who came to the Cape had influence with the government and worked for the freedom of the Africans. In 1807 the slave trade was ended at the Cape, and a labor shortage developed once again. Caledon, the first governor during this period of British occupation of the Cape, and his successor, Cradock, issued certain restrictive proclamations concerning the Hottentot population. These proclamations were made in an effort to ameliorate the deteriorating labor situation without antagonizing the missionaries. Among other things, these proclamations required any Hottentot who wished to move from one of the estab-

lished Districts to another to have a pass. Restrictive pass
laws suited the Africans no better then than in the 20th
century, and the Hottentots went to live across the borders
of the Colony and at mission stations. It was not long
before the Africans had a champion. He was Dr. John
Philip, a superintendent of the very active London Missionary Society. He arrived at the Cape in 1819, and by
1828 he had forced Governor Somerset to repeal the
proclamations made by Caledon and Cradock and actually
succeeded in obtaining for the Hottentots a measure of
equality with the whites. It is curious to find that Philip,
in addition to being a champion of racial equality, was also
a confirmed segregationist. In his writings on the frontier
and from his actions there, it becomes clear that he felt
that spatial separation of the races was a matter of self-
protection for the more backward Africans. He worked
as vigorously for the termination of frontier expansion
as he had for the equality of the Africans living in the
settled Cape Colony, but with less success.

Although the early years of British occupancy at the
Cape were prosperous years economically, the labor shortage notwithstanding, the Cape did not escape the effects of
the great depression of the 1820's in Europe. Coinciden-
tally with the onset of the depression, toward the end of
the second decade of the 19th century, Britons other than
administrators and missionaries began to move to the
south of Africa. The first group of settlers of this kind to
come from the British Isles was a body of about 400 Scots
who arrived in 1817. The local administration approved of
this increased rate of immigration, and the home government began to aid it by making financial grants to those
who wished to go. An estimated 5,000 British settlers arrived at the Cape in 1820, and many were sent out to the
area of Port Elizabeth, where they received a land grant.
Port Elizabeth stands on Algoa Bay, which was sighted
and named by the early Portuguese navigators, and it is not
very far from the place where the Great Fish River reaches
the sea. The settlement at Port Elizabeth and its hinter-

land became a small British colony, and many of the place names in this area originated in this period of settlement.

One of the important changes that had taken place with the surrender of the Cape to Britain was in the form of administration. The Policy Council ceased to exist, the Governor was the sole and supreme power, and he received only occasional directions from Britain. Hence, the Cape was ruled without any mechanism for local participation. One of the conditions against which many local settlers had agitated under the Dutch was the lack of sufficient participation in decision-making; now, there was not even a semblance of such participation. It was not until 1825 that a small concession was made in the establishment of an Advisory Council. This body of local officials could offer the Governor advice, but suggestions were not binding. Nevertheless, local people did again have an official sounding board, and it was considered a good omen for the future. A major step took place in 1834 toward representative government and was the result of continued pressure in the colony for political advancement. This involved the replacement of the Advisory Council by a Legislative Council, which sat government officials next to appointed representatives of the settlers. This was the first real move in the development of local, elected, representative government, eventually secured by the colonists in 1854.

The first 30 years of British occupation led to a great exodus of settlers, who moved into the interior with such belongings as they could transport in their covered wagons. After a century and a half of Dutch rule, the British appeared as "outsiders" to many settlers, and the British acted harshly in their efforts to anglicize the Cape. Governor Somerset especially was set upon this policy, and in 1822 he proclaimed that English was to become, in 1827, the only official language in the Colony. The proclamation took effect in 1828, when it coincided with the initiation of an entire new judicial system. This replacement of the old system of administering justice by a new judiciary was necessary, but it was unfortunate that the two changes—

linguistic and judicial—took place at the same time. The effect of the language proclamation was, of course, resentment among the Dutch-speaking majority. The judicial changes affected particularly those in the border regions of the Districts, where people were not used to legal complications when white-black, master-servant relations were concerned.

Other changes were taking place in education and religion. Schools were established by Somerset, and teachers were brought from Scotland. Education in the Colony was a problem as it is in South Africa today. Quarreling between Somerset and certain of the teachers led to the establishment of two rival high schools in Cape Town, which merged after Somerset's departure and were the beginnings of the University of Cape Town, an English-language institution. In addition to teachers, Scottish ministers came to the Cape to join local congregations of the Dutch Reformed Church, which was strong in the Cape then as it became throughout South Africa. The religious changes were not so severe as were some others imposed during the first three decades of British rule. Many of the Scottish ministers had acquired some knowledge of Dutch while spending a preparatory year in Holland, and they attempted to use and perfect this asset rather than force the English language upon their congregations.

It is possible that the change which caused the most settlers to leave the British-dominated area for good was wrought not at the Cape but in England, in 1833. This is the year during which the campaigns against slavery came to a climax with the decision of Parliament to end the practice throughout the Empire. Some 30,000 slaves in the Colony were affected by this decision. Slaves were worth money, and at the Cape their labor was at a premium. Although some compensation was paid to the farmers, many lost heavily. The hostility of the frontiersmen reached a new peak.

Life otherwise on the frontier, remote from the political and social troubles of Cape Town and its immediate sur-

roundings, changed but little during the first 30 years of
British rule at the Cape. Naturally, there was friction
with the new rulers; but the settlers were more immediately concerned with the Xosa, and three major wars were
fought between white and African: in 1811, 1819, and
1834. Each time, efforts at establishing a permanent negotiated peace and mutually acceptable boundaries failed,
and the British government, like the Dutch before it, did
not have the manpower and resources to protect the frontier adequately. In addition, it did not have the will: it
blamed the frontiersmen for the conflicts and sought to
establish treaties with the African chiefs. To make matters
worse, a series of great droughts affected the disputed
lands between 1831 and 1834, which made every inch of
pasture seem more important and which sent white and
African cattlemen in new directions to seek grass and
water. Meanwhile, the British settlement at Port Elizabeth, had established itself, and trade was in progress
between African and white, causing further contact and
friction. The hinterland of Port Elizabeth was in itself a
frontier, and, like the frontiers of Graaff Reinet, it was
outside the effective control of the government at the Cape.
Further westward, some white cattlemen were crossing
the Orange River and making contact with another African people, the Griqua. They were buying land from the
Griqua and were more remote from the Cape than ever.
Their reports of good land available even farther north
reached farmers along all the frontier and in all the Districts.

What happened in the Cape Colony among the whites
during these first 30 years of British rule became of far
greater importance to the course of events during the next
century than were the wars on the frontier of the settlement. Had harmony prevailed at the Cape, a very different kind of expansion might have resulted. As it was,
disunity ran rife in many aspects of life. The Dutch settlers
were harassed by legal, educational, linguistic, and religious changes, and Dutch settlers at the Cape, particularly

in the Districts, did not have much experience with, and did not take very kindly to, change. Perhaps the decision on the part of the British to abolish slavery was an important factor in this picture of division. Certainly the settlers, especially those on the frontier, must have felt that a government which imposed such unpalatable changes without affording even the minimum of protection was no asset. These considerations and the reports from the north of good lands, open and ready for occupation, caused many of the settlers, in 1835 and 1836, to break with Cape Colony and trek into the interior. And so the period of gradual expansion of the frontier was superseded by a spearhead invasion of the deep distances of the subcontinent—the "Great Trek"—by groups of pioneers who took with them strong religious beliefs, a determination to remain free, and very little in the way of personal belongings.

| *"Empty Space"*

There was a particular and most important reason for the reports which came from the farmers who had entered Griqualand and the adventurers who had penetrated far into the South African plateau. The interior, they reported, contained vast, empty tracts of fertile land teeming with game. Elsewhere, the whites had encountered Bushmen, Hottentots, Xosa, and Griqua, and a great migration of Bantu peoples southward had been in progress for centuries and was still going on. Yet, said the scouting parties, this rich, grassy land was uninhabited. Was it? Why? The answer has much to do with the outcome of the Great Trek. It accounts in part for land claims still made in South Africa.

The great Bantu migration to the south *had* been in progress over several centuries, and it is probable that the African population of the subcontinent in the early part of the 19th century exceeded 5 million. Like the Xosa with whom the settlers had come in contact, many of these

Bantu peoples were pastoralists, and some were beginning
to show tendencies toward sedentary agriculture. The opinion is commonly held that Africa before the coming of the
white man was the scene of almost universal war between
African peoples, but in fact this was not so. Although there
was friction in some regions at certain times, a state of
constant war did not prevail. It is true that excessive
fragmentation and lack of cohesion can be called a major
cause for the success of the slave trade, which went on,
unimpeded, for centuries; there simply was not the organization and unity on the part of the Africans to prevent it,
let alone the necessary weapons. But although the indigenous population of Africa was divided, racially and
otherwise, war was not constant and omnipresent between
the various groups which were making their way from
central Africa into the south.

The Bantu movement to the south was largely a search
for land. It appears to be anomalous, then, that the white
trekkers who ventured into the interior should encounter
empty spaces; the Bantu had reached the eastern Cape well
before the beginning of the 18th century, and the interior
must have been under the influence of the immigration for
a long time. Several bloody border wars occurred along
the Great Fish River between 1800 and 1836, and these
wars were not merely the result of interracial dislike but
had as their cause a need for land on the part of black as
well as white. Thus there is evidence of population pressure in one area of the southernmost part of Africa,
while not many hundreds of miles distant, lands of quite
high fertility went unoccupied.

The Xosa and the frontiersmen fought battles not just
because in their respective migrations they happened to
meet. In effect, the Xosa were being driven toward the
south and southwest at just the same time as the frontiersmen moved farther and farther away from the hostile
government at the Cape. The same forces that were driving the Xosa ahead of them were the cause of the empty
spaces of the interior. Black Southern Africa beyond the

Cape borders was at war. The Xosa were fleeing, not migrating, and the empty spaces were not permanently empty. It was sheer coincidence that the trekkers found the interior of South Africa empty—a coincidence which present-day Afrikaners disregard when they give a so-called moral justification for South African policies concerning living space for the various racial groups.

If war was not as prevalent in Africa as might be thought, then why was there major war between African peoples in the early 1800's? From whom were the Xosa fleeing? From the Zulu, who in the first two decades of the 19th century gained rapidly in strength and numbers as well as organization. With strong leadership and rigorous discipline, the Zulu became the most powerful people in the south of Africa, and in the 1820's they set out on a course of conquest and devastation. Their armies trained constantly in battle tactics, knew no defeat, and scattered before them in all directions the African peoples who had been living anywhere near. Battles raged along the coastal slopes of the Drakensberg, along the escarpment and onto the plateau, and the losers were driven as far afield as the western Transvaal, across the Vaal River, southern Rhodesia across the Limpopo River, and Basutoland's high plateaus. The Xosa were driven south and southwest and were repelled on the borders of the Cape Colony. The Africans who had been driven out of their original home to other areas often came into conflict among themselves in these regions, so that the wars of the Zulu reverberated far and wide throughout the subcontinent. In certain places, fleeing Africans were united by strong leadership, as shown, from this period on, by records of the Matabele in Rhodesia and the Basuto in Basutoland. These Basuto are particularly interesting. They owe their survival and even, in many ways, their present-day semi-independence to their great leader Moshesh, who was a man of extraordinary intelligence and resourcefulness. In his long reign, Moshesh dealt first with the Zulu threat and later with

the British to avoid incorporation of his small country into
a larger state.

| *The Great Trek and the Republics*

Dissatisfied with policies at the Cape, attracted by the
promised lands of the plateau, bent upon maintaining their
national character and religious convictions, a number of
the cattlemen—Boers (pronounced *boor*, not *bore*)—
pushed to the interior in 1836. This year marks the beginning of a new phase in the history of settlement in
Southern Africa. Until this year, there had been one Cape
Colony, whether or not it was a divided settlement. There
was one government and one official ruler: Britain. The
Great Trek was aimed at the establishment, in the interior,
of Boer republics, free of British domination and free to
practice religion and education in the Dutch language.
Here slavery would not be prohibited. These republics
were to be farming, cattle-ranching republics, and the possibility of mineral wealth to be found in the interior probably never even entered the thoughts of the great majority
of the trekkers.

Perhaps it is advisable here to recollect that Dutch had,
for some time, been only an unofficial language in the Cape
settlement and that contact with Holland, for obvious
reasons, was now at a minimum. Hence the Dutch language spoken here had begun to change, and there was a
certain amount of mixture with other languages. Malay,
Hottentot, French, and English words had become a part
of the new language, which was also changing structurally.
In addition, the outlook of the people was no longer really
Dutch. Holland began to seem more and more remote to
the settlers on the frontier and the Boers who trekked into
the interior, and the idea of being Dutch and oppressed
by the British began to make way for the spirit of being
South African and ruled by foreigners. Here is another
of the important strands that runs through all South Afri-

can history until the present day. The new language that was developing through isolation and exposure came to be called Afrikaans, and although the term "Dutch" was applied to the Boers for many years after the Great Trek, it is perhaps best to refer to them henceforth first as Boers, then as Afrikaners. The word "boer" means "farmer."

In the Great Trek, the Afrikaners found some of their great leaders. They had long lived a kind of nomadic pastoral existence as they moved eastward and northward through the years, pushed the frontier ahead, and sought pastures during the droughts. Yet this Great Trek involved different and more challenging hardships. A body of trekkers, perhaps two hundred in number, loaded all they could carry on their wagons, drove their animals ahead of them, and severed all contact with the Cape Colony. Although the interior was largely empty, there were nevertheless bands of Bushmen who could attack at any moment, and there were African peoples along the fringes of the central plateau and in isolated areas within the interior, so that contact and conflict were always possible.

The Great Trek took place largely from the eastern Cape to the north, and it took the Boers from the slopes of the east-west trending parallel Cape Ranges, along which they had been moving, to the Great Escarpment, as forbidding here as anywhere. In order to reach the plateau, they had to ascend the escarpment. Their animals had to be driven up very steep slopes, and many were lost. Wagons had to be taken apart and carried up piecemeal. The Boers did not lack determination, but there were greater dangers in store for them. Because of the relative emptiness of the land they entered, the parties began to break up and scatter. For this the Boers paid dearly, since they were attacked by Matabele who killed all members of some of the small units before the trekkers decided to unite once again. On the plateau, one of the decisive battles came late in 1836 when the trekkers and the Matabele clashed in great numbers near the Vaal River,

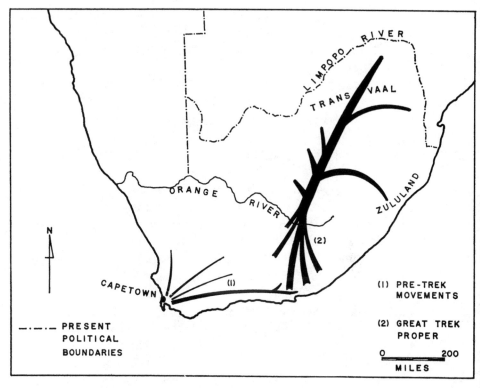

MAP 3   *The Great Trek*

one of the northern tributaries of the Orange. The Boers survived, and their domination of the plateau had begun.

Once in the interior, the trekkers had the choice to cross the Vaal River and go into the reputedly more fertile north ("Trans Vaal") or to attempt to establish a link between the plateau and the sea by entering Natal, also known to be fertile and desirable. Some parties went to the north and actually reached the foot of the Zoutpansberg, today the northernmost range in South Africa in the northern bend of the Limpopo River, and others turned eastward with the intention of settling in Natal. This movement eastward proved to be a fateful one, because the Boers who entered Natal moved straight toward the lion's den.

Going into Natal, of course, meant descending the Great Escarpment. The Boers were well aware of the Zulu's presence and power and remained on the edge of the plateau while their leader, Retief, rode toward Durban and on to the Zulu king, Dingaan, with intentions to negotiate for land. The negotiations were quite successful, and it seemed that a treaty might be concluded with the Zulu. Unfortunately, the trekkers who had remained on the edge of the plateau above the escarpment to await the results of the talks with Dingaan became impatient. When Retief sent news of his successful negotiations to the waiting Boers with instructions that they await his return so that he might personally lead them into Natal's lowlands, the trekkers could wait no longer. They began to descend the escarpment. Now Retief had signed an agreement with Dingaan to the effect that he, Retief, would recapture for the Zulu king a number of cattle which had been stolen by a rival tribe. This was in return for the land to be given to the Boers. For the Boers, returning those cattle to Dingaan seemed a relatively easy matter, but it was to be done before anyone entered Natal, and this was why Retief told the Boers to wait for his return.

The premature movement of the Boers into the Natal lowlands had disastrous consequences. Upon his return to the Zulu king's capital, Retief, unaware of what was happening along the escarpment, was killed with all seventy of his party, on orders of Dingaan, who had been informed by his scouts that the movement of the Boers had begun. This occurred early in 1838, scarcely two years after the Great Trek had begun, and 1838 was one of the hardest years for the Boer trekkers. The Zulu, having killed Retief and other Boer leaders, deployed their warriors against the unsuspecting Boers who were entering Natal, and whole units were completely destroyed. Other leaders on the plateau heard of the Natal trekkers' plight and came to their aid but lost some crucial battles against the Zulu. Hundreds of trekkers died, and thousands of cattle were taken by the Africans. In part, these losses resulted from division among the Boers themselves, whose leaders competed for prominence. Whatever losses they suffered, however, the Boers were not persuaded to leave Natal. It was late in the year 1838 when the Boer leaders Cilliers and Pretorius, with about 500 men, fought the masses of the Zulu at Blood River—and won. This, the Battle of Blood River, is still commemorated by a Thanksgiving Day every December 16. The battle was the beginning of the end of Zulu power in Natal. Dingaan survived, but only to be defeated by an army consisting of Boers and Zulu who had deserted Dingaan and joined his half-brother, Panda. So now it was the Zulu who were divided, and this led to their defeat. Panda, in return for the aid of the Boers in defeating Dingaan, agreed to hold to the treaty established between Retief and Dingaan.

The war between the Zulu under Dingaan and the Boer-Zulu army of Panda had one additional and very important consequence. At the Cape and in Britain, the British were beginning to view with increasing alarm the movement of the trekkers into the interior, their troubles there, and the amount of land they were claiming. Hence when the Boers were engaging the Zulu at Blood River, a British army was on its way to occupy Port Natal (now Durban), after which efforts were to be made to end hostilities between Boer and African before unfavorable consequences were felt on the frontiers of the Cape Colony. The British did occupy Port Natal, and they were in the process of helping the Boers and the Zulu settle their differences by talks held in the town when the Zulu were fragmented by the Dingaan-Panda struggle. Renewed war meant failure of the British mission, and since the British had no orders to take all of Natal, they left the region. Natal, from the Drakensberg to the coast, was in the hands of the trekkers, just four years after the first party of trekkers had left the Cape Colony. This was the first, if short-lived, of the republics which the Boers were to establish in various parts of Southern Africa. The government at the Cape reacted by proclaiming that even trekkers who left the Colony would remain British subjects. There was concern over the outflow of Boers, and efforts were being made to keep them from going.

This concern was among the factors which caused the Natal Boer Republic to be short-lived. The British worried not only about the depletion of population at the Cape and the loss of control over the people who were departing but also about the effect on British interests elsewhere in Southern Africa. Once the Boers took possession of Port Natal, ships began to bypass Cape Town and call there. Needless to say, the ships calling at Port Natal were mainly Dutch, but the British administrators feared that Port Natal might grow to challenge Cape Town's position of dominance in overseas trade.

Within the Natal Republic itself, things were far from stable. The frontiersmen who had fought the Xosa and the Zulu were used to taking what they desired; they were not used to obeying a government. They showed remarkably little interest in establishing a government at all, and with the common enemy, the Zulu, defeated, old rivalries again made their appearance. Slavery in an only slightly modified form persisted. Along the agreed borders with the Zulu, conditions began to resemble again those which had existed along the Great Fish River and elsewhere on the Cape frontier. Cattle raids, land grabs, and minor skirmishes occurred frequently, and the Volksraad (People's Council) seated in Pietermaritzburg was as powerless to stop them as the Cape government had been before. Battles fought on the southern boundary with the Africans again raised fears that there would be reverberations at the Cape Colony's eastern borders. The British decided to invade the Republic of Natal.

Boer and Briton thus for the first time found themselves locked in combat. When the British landed at Port Natal in 1842, the Boers defeated them in battle, but reinforcements from Cape Town and Port Elizabeth soon enabled the British to stand their ground. A period of uncertainty followed the fighting, which ended in a standoff, with the Boers hoping for help from other trekkers in the interior and from Holland, the British waiting for London to make its decision regarding the future of the Republic, and Africans re-entering Natal from north and south. Eventually, in 1845, after two years of negotiating, the Republic of Natal became another British colony. The division among the Boers on the question was again apparent. Many left Natal and returned to the plateau once again, escaping for the second time the British rule they had sought to escape previously when they left the Cape. Others stayed and became British subjects. They were tired of trekking, and some came to realize that attempts at escape from the British in Southern Africa were not likely to be successful.

With the coastline now entirely in British hands, the Afrikaners found themselves isolated on the interior plateau, and attention is now focused once again upon those high grasslands north of the Orange River and west of the Drakensberg and the Great Escarpment. In the 1840's, various attempts were made in this region by the Afrikaners to establish republics. In fact, several temporary Boer republics did develop. Such present South African towns as Winburg and Potchefstroom over a century ago were important centers of the embryo republics of the highveld. However, it would be false to suggest that there was any more success in the efforts to establish stable government on the highveld than there was in Natal. The trekkers refused to submit to government, land-grabbing proceeded as elsewhere, conflicts arose in increasing number and intensity with African peoples. A battle between Boers and Griqua led to the intervention of the British as had the battle between Boers and Zulu in Natal, but this time the British did not stay to form a colony. A change of policy in London combined with Boer resistance to effect British retreat from the highveld to the old border of Cape Colony, the Orange River. By 1854, the Boers found themselves free again to determine their own form of government, and in some ways the Great Trek had accomplished its original purpose. A great deal of jealousy and personal rivalry remained, and the interior was fragmented on occasion into five small republics. However, with the departure of the British some semblance of stability appeared, and two Boer republics were formed which were to play important roles in the fifty years of their survival. Today, areas roughly corresponding to these republics remain as provinces of South Africa. In the south was the Orange Free State, and north, across the Vaal River, was the South African Republic, now known as the Transvaal.

The history of the city of Pretoria provides some indication of the nature of these republics. Pretoria, now South Africa's administrative capital, is one of the larger cities in the subcontinent. It has retained its early importance and is growing rapidly today. With the establishment of the South African Republic, Potchefstroom, founded in 1839, was incorporated. In 1856, Potchefstroom is estimated to have had some 2,000 white inhabitants. The annexation of Natal had caused the return to the Transvaal of a large number of trekkers, and many of them stayed in the beautiful eastern section, where the town of Lydenburg became the center of population. At the foot of the Zoutpansberg, in the far north, was the town of Zoutpansberg, with perhaps 200 white people, and approximately in the south-central Transvaal was Rustenburg. Each of these towns was the focal point for an area of relatively dense white settlement, and really the South African Republic consisted of four "kontrys" or individual colonies. There was some argument about the proper location for the Volksraad of the new Republic, and the leaders eventually decided to create a new capital, centrally located and protected by the surrounding settlements. However, the townspeople of Lydenburg—and this is an indication of the relative trivialities that could cause serious disruption—upon receipt of the news of the chosen location declared that the territory of Lydenburg would secede from the Republic. Lydenburg had previously objected to the whole system of government in the Republic, was against the constitution, against the idea of a president and central government, and opposed the building of any new capital. To avoid the secession of Lydenburg, the plan for Pretoria was shelved, and Potchefstroom remained the temporary capital.

Pretoria, however, still had its supporters, and the site chosen was a good one, on the southern flanks of the Magaliesberg range. Even without the impetus of capital status, a hamlet developed, and slowly the settlement enlarged. Someone by the name of Du Toit, with the most primitive of surveying equipment, drew a town plan, still visible in the urban pattern today. In succession, an administrator-magistrate, doctor, and first storekeeper settled in 1857 and 1858. In 1858 Pretoria consisted of seven or eight scattered, thatch-roofed buildings, but by 1860 there were some 80 families. In 1861 came the first minister, soon after Pretoria was finally made the permanent capital of the Republic. It was not until 1868 that a government building was constructed, although the government press was moved to the capital in 1863. Even at this late date, few Afrikaners cared whether their government moved fast or slow, here or there. In 1869 a President of the Republic, when he was about to be sworn in, objected that the government was in Pretoria and his house in Potchefstroom: he wanted the government moved back to Potchefstroom so that he might live in his house!

Besides apathy and trivial disputes, the Republic faced grave problems. Money was scarce, for one thing. Since 1857, "mandates," or government-debt certificates, had been issued. Government officials put these out at will, and by 1864 there were fears about the growth of the national debt. Meanwhile, administrators and other officials in the government were themselves being paid in these certificates, with which they could, of course, buy nothing. In 1865, government people began to threaten with resignation if they continued to be paid in mandates, and in 1866 the first pound notes were issued, after an abortive attempt to establish a Dutch-type currency system. Almost immediately after its first issue, the money began to decline in value, and again there was need to resort to debt-certificates. The economic situation of the Republic was so chaotic that in 1872 the government sought a loan from the Cape Commercial Bank to the value of 60,000 British pounds sterling. In 1873 this loan became reality, and a branch of the Cape Commercial Bank was established at Pretoria. In addition, gold was being discovered in various parts of the Republic, though not yet at the Witwatersrand where Johannesburg is now located, and the faltering economy began to change for the better. Pretoria, about this time, had a white population of over 3,000.

The Orange Free State Republic, partly by virtue of location, had a somewhat different history. The Afrikaners in the Free State were not as far from "civilization" as those in the South African Republic, and they maintained contact with the Cape Colony. Moreover, they had the strong Basuto under Moshesh to their east, a matter for constant concern. While there was tension and even civil war in the South African Republic, the Free Staters were united to some extent against a common danger. Nevertheless, government in the Orange Free State Republic was barely more effective than it was in the South African Republic, and personal rivalries did much harm. All efforts at unification of the two Republics failed. Bloemfontein, its partisans were quick to point out, had been founded and was the capital of the Free State before the argument in the South African Republic over Pretoria even started. The Orange Free State profited somewhat by the activity of traders from the south, but there was no organized economy or great wealth.

| Diamonds and Gold

The Boers had trekked into the interior to acquire land and to escape British rule. Their republics survived for

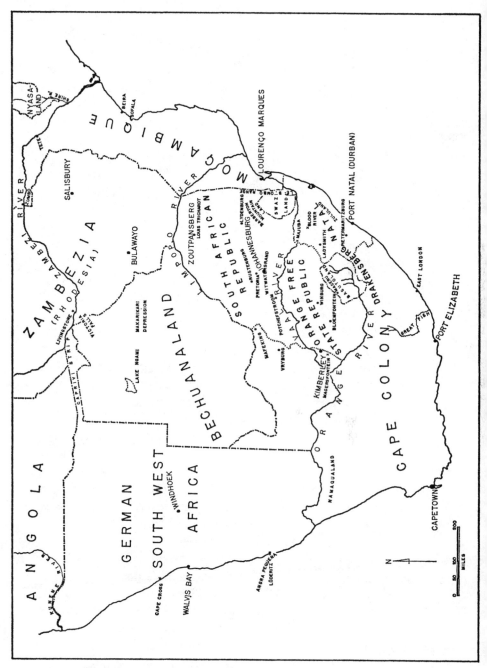

MAP 4 *Southern Africa: Evolution of the Political Framework.*

about a half-century, always troubled by internal friction, battles with the Africans, conditions of slavery, and eventually they came into serious conflict with the British. This half-century saw the discovery within the republics of two mineral deposits which attained world fame and which for some time formed the major basis of the South African economy.

Diamonds were discovered along the Vaal River, not far from where it joins the Orange, and along the Orange itself. In 1867 the area where the diamonds were found had no well-defined political status. This region lies close to the eastern edge of the Kalahari Desert, and it was then, as now, arid and barren. Precisely for this reason, it was one of the regions into which Africans, in this case the Griqua, had been driven. The accidental discovery of the diamonds created disputes over the ownership of the land, while diamond-diggers by the hundreds began to move in. The Cape Colony, the Orange Free State, and the South African Republic were all involved; as fate had it, the land was located in such a manner that all three could claim part or all of it. Meanwhile Chief Waterboer of the Griquas insisted that this was his people's land. The President of the Orange Free State sent law officers to maintain order among the unruly diggers, and the President of the South African Republic produced records of treaties with Africans in the area to support his claims. The Cape Colony saw the usefulness of the land but could not without consent from Britain enter the area. To cap the confusion, the diamond-diggers proclaimed themselves independent. Thus, for a short time, there was on the map of Southern Africa a Diggers Republic.

The diamond fields were soon world famous, and men came from many countries to search for wealth. Kimberley became the diamond center and the fastest growing town in Africa. By 1872, over a million dollars worth of diamonds had been found, and the population of the area was to be counted in the tens of thousands, black and white. This is where the African in southern Africa first became an industrialized wage-earner. Africans flocked to the mines from many parts of the subcontinent.

The diamond rush ended the economic backwardness of the interior republics and reawakened activity at the Cape and in Natal. It also caused bitter disputes over land, not between white and African this time, but between Boer and Briton. One of the services most urgently required in the subcontinent was communication. Now, money became available for the building of railroads from the coastal ports to the interior. In this respect, South Africa's relatively early start has had very impressive results. Today, South Africa is the only country in Africa with a really extensive network of railroads, and this network was begun with the money the diamond industry provided. By 1885, Kimberley was linked by rail with Cape Town, Port Elizabeth, and East London. The diamonds were in the interior, however, and the ports were on the British-held coast. The British schemed to take control at both ends of the new railroads. A so-called arbitrator from Natal settled the Free State–S.A. Republic–Waterboer dispute over the diamond fields in favor of Waterboer. Waterboer had previously asked that his people be put under British jurisdiction, and with the award of the lands to Waterboer, the diamond fields became British land and were made part of Cape Colony. The Orange Free State never agreed to the arbitrator in the first place and did not accept the obviously biased decision. Britain paid the Free State an indemnity when there were objections from its government about the procedure. But the rich lands had been taken. It was a sign of things to come.

Among the men drawn to the diamond fields was one who came to alter the course of history. The mind of Cecil John Rhodes was set upon the expansion of British influence throughout Africa, and the money of the diamond fields gave him his opportunity. Rhodes was one of the prominent men in the diamond business, since he, unlike so many others who came to Kimberley, was prosperous before he arrived. In time he succeeded in establishing one of the most famous South African mining companies: De Beers Consolidated Mines, in which he was a major shareholder. Rhodes had his eyes on the area west of the South African Republic, Bechuanaland, and the area north of the Limpopo River, Zambezia. One of Southern Africa's peculiarities is a legacy of Rhodes's imperialistic expansionism. The railroad from South Africa to the Central African Federation runs not through the Transvaal but through Bechuanaland, and railroad travel to Salisbury from Johannesburg is still cumbersome today because of this arrangement.

While all eyes were on the diamond rush in the interior, a less-heralded event was taking place in the 1860's in Natal, an event destined to have great importance in later years. Natal Colony had been found to be extremely fertile, particularly in the moist, warm lowlands. The soil and climate proved favorable for the growing of sugar, and large sugar plantations began to develop. The Zulu, however, and the other Africans in the region, did not constitute a satisfactory labor force. It was decided to import Indians. In 1860 arrived the first ship with indentured Indian Laborers, who were to stay for a period of three years, then to return, if they desired, to India. The majority of the Indians, after serving their period of labor on the sugar plantations, preferred to stay in Natal and did, becoming family servants, vegetable farmers, or traders. These Indians were remarkably skillful, and many attained wealth in South Africa.

Although gold had been discovered by an explorer in the region of Southern Rhodesia, lands beyond the South African Republic were still virtually unknown even at the time of the diamond rush. However, an increasing number of missionaries, hunters and traders were going north, and among them was Dr. David Livingstone. He arrived in Cape Town in 1841, with plans to travel through the interior, and he set out at once to Bechuanaland. Livingstone made several important discoveries, some of which came to the attention of Rhodes and furthered his desire to take possession of these regions for Britain. In 1849, Livingstone discovered Lake Ngami, part of the Makari-kari Basin in Bechuanaland, and in 1850 he first saw the Zambezi River above the great falls. He turned west and reached Luanda overland. In 1855, he discovered the Victoria Falls and proceeded down the Zambezi River to its mouth. Next, he traveled up the Shire River, which connects Lake Nyasa with the Zambezi River, and observed the slave trade still going on there. His reports partly influenced Britain to take control of Nyasaland. Dr. Livingstone's fame and the association of his name with those of Mungo Park and other early explorers make it difficult to realize that some of his most important discoveries were announced barely 100 years ago.

Rhodes, not the discovery of gold, started the major movement of whites into what is today Southern Rhodesia. King Lobengula of the Matabele granted a major concession of land to Rhodes, and it was 1890 before the Pioneer Column of less than 200 white settlers departed from Rhodes's farm near Kimberley on its way to Zambezia, as the territory was then called. This was after the region had been declared a British sphere of influence, which had come about in 1888, thanks to Rhodes's efforts in convincing the British High Commissioner to negotiate a treaty with Lobengula, which in turn had led to the formation of the British South Africa Company.

Whereas the Free State was involved in the discovery of diamonds, it was gold which changed the course of events in the South African Republic. At the time of the first big strike, 1884, the Republic was just getting back to normal after an effort on the part of the British to annex it. The Boers had fought the British in 1877 and secured a truce. In 1881 independence of the Republic had been re-established, and in 1883 its strongest leader, President Kruger, had been sworn in. Like the diamonds of Griqualand, the gold of the world's most famous gold field, the Witwatersrand, was found by accident. In 1884, two brothers were digging on the Witwatersrand (Ridge of White Waters) the foundation for a small house. They struck gold, though not the main reef. In 1886 the major outcrop of gold-bearing reef was located, and the boom was on.

This time, there was no doubt about the jurisdiction over the mineral deposits. The Witwatersrand lies in the south central Transvaal, only slightly more than 30 miles south of Pretoria, then the capital of the young republic. The discovery was of such magnitude that a new rush began, less than 20 years after the diamond rush. The gold rush involved many more people than the diamond rush had done, and again there was a wave of immigrants from all parts of the world. Within a very short time after the first discovery, the "Golden City," Johannesburg, began its unequaled growth. The ideal "farming republic" of the Boers was endangered by tens of thousands of "outlanders" who paid little heed to local laws. Johannesburg was swamped with miners, speculators, confidence men, traders, and adventurers. It is not surprising that the government at Pretoria could not cope with the Witwatersrand. To keep what power they could, the Boers refused to grant Johannesburg its own municipal government. Hence the town grew chaotically and without proper planning, and order was not maintained.

Like diamonds and Kimberley, gold and Johannesburg stimulated railroad development. Johannesburg was soon a much larger city than Kimberley was at the height of diamond-digging activity, and the coastal ports began to vie for the first railway connection with this rich new hinterland. By 1892, only six years after the major strike on the Witwatersrand, Johannesburg was linked by rail to Cape Town, Port Elizabeth, and East London. Eventually, connections with Durban and Lourenço Marques were also established, and the network of South African railroads began to take shape. The isolation of the interior republics had come to an end, and their most difficult times were about to start.

The lack of effective government in the South African Republic, except when it came to discriminating against foreigners, gave rise to a considerable amount of agitation in favor of British annexation or, at least, a termination of the prevalent Pretoria policies. As elsewhere in Southern Africa, British empire-builders were at work in Johannesburg, and friction between foreigner and Afrikaner promoted their cause. The situation in the Republic after the discovery of the gold deteriorated steadily. One of the figures of importance in these events was Rhodes. Rhodes had seen the British flag raised by his ally Jameson over Salisbury, Zambezia, in 1890, and he had been instrumental in the annexation of southern Bechuanaland to the Cape Colony and the establishment of a British protectorate over the northern half of that territory. A railroad was under construction from Kimberley to Mafeking and on to Bulawayo and Salisbury in Zambezia, and several of Rhodes's dreams were materializing. Also, Rhodes had attained a position of political power as Prime Minister of the Cape, and with his financial and political powers combined he was working from a position of great strength. Rhodes saw the tremendous wealth of the South African Republic and thought about the possibility of federation between his colony and the interior republics. However, the attempt at British annexation of 1877 was still fresh in Boer minds, and the South African Republic was under the guidance of President Kruger, who was unalterably opposed to any federation with the British.

Kruger, with his unyielding Calvinist nature, was not disposed to give in to Rhodes the urbane imperialist. In Johannesburg, the disgruntled citizens set up a Reform Committee to work for concessions from the Afrikaner government. It was singularly unsuccessful, and when negotiations failed, another plan was drawn up. Rhodes and Jameson, with the aid of the Committee, plotted an invasion of the Republic and justified it with the claim that internal friction in the Republic endangered peace and lives. In this case, there was less justification than there had been in Natal in 1842 and in Transvaal in 1877, and the invasion, the "Jameson Raid," as the attempt at the overthrow of the Boers came to be called, was not well planned. Jameson entered the South African Republic from Bechuanaland on December 31, 1895. Like Captain Smith at Port Natal in the 1840's and General Colley at Majuba in 1880, Jameson underestimated the Boers in battle. At Doornkop, he suffered one of the several inglorious defeats of the British in Southern Africa during the 19th century.

The Boers, having captured the invaders, now witnessed a sample of British justice. Kruger tried a handful of the leaders of the insurrection, and sentences were extremely light; no one was put to death, and some were punished with only a fine, which friends of Rhodes should have had no trouble in paying. In a frontier country, this was mild sentencing. Jameson himself was sent to Britain for trial, a wise decision on the part of the Pretoria government. Britain, which had consistently denied all complicity in the Raid, proceeded to give light prison sentences to the major offenders in the matter. Jameson served less than five months in jail before being freed on grounds of "ill health," and no one in the British government resigned or showed the slightest regret. The immediate consequence of this was the signing of a military treaty between the South African Republic and the originally somewhat more moderate Orange Free State, which came to the support of the Transvaal as a result of indignation over these matters.

Britons in Britain and Africa did not take well to setbacks, and British armies began to mass on the borders of the South African Republic, while the republicans also armed themselves. Repeated negotiations between the conflicting parties took place, without positive results. At the last of the prewar conferences between Britain and the South African Republic, held in Bloemfontein in 1899, the latter demanded that Britain remove its troops on the borders of the republic. Britain refused, and on October 11, 1899, war between Britain and the tiny republic broke out.

| *The Boer War*

In the Great Trek, the Afrikaners had found inspired leaders, and the Boer War likewise threw into national prominence several men who first led the Boers in combat and, after the war, proceeded to dominate the South African political scene for nearly half a century. The forces of the Orange Free State, which honored its treaty, and of the South African Republic were put under the command of General Louis Botha. Two other generals played significant roles. They were Jan Smuts and James Hertzog. Botha, Smuts, and Hertzog were each to be Prime Minister of their country, and Smuts became one of the world's great statesmen.

The Boer War started with such reverses for the British that the Boers were able to invade Natal and the Cape. The British commanders Buller and Gatacre were defeated by Boer commandos, and at Magersfontein, Lord Methuen's army was eradicated. Many experts have written on the failure of Boer strategy after these early weeks of triumph. Instead of penetrating to the coast and gaining a hold over the communication system, the Boers spent their energy on sieges of Ladysmith in Natal and Mafeking and Kimberley in Cape Colony. These tactics gave the British time to reorganize their defeated troops, to rush in additional men and materials from Britain and India, and to appoint Lord Roberts and Kitchener as the new commanders. In 1890 the war (not "over by Christmas," as had been forecast the previous year in England) was being carried to the Boers, the sieges were broken, and Bloemfontein and Pretoria were taken. With the cities and towns taken, the second phase of the Boer War began, in the form of a guerilla struggle which lasted nearly two years. The British, while attempting to remove all possible hiding places for the Boer commandos who harassed transportation lines and sabotaged everywhere, took to burning down farms and herding women and children into concentration camps. Here they died by the thousands, a costly matter for the whole population of the subcontinent and a tragedy which has retained its significance in South African nationalism until the present day. Numerous prisoners were also packed off to such places as Bermuda, St. Helena, and Ceylon.

The hopeless war dragged on, partly, because the Boers hoped that some foreign power would intervene on their behalf. Several times, Kitchener made efforts to persuade the Boers to surrender. His first offer was made as early as March, 1901, but although the Boers were in a very poor state, they carried on the fighting. Even when peace came on May 31, 1902, with the signing of the Treaty of Vereeniging, there were still districts where the Boers were willing to carry the war on, though the majority of the republicans admitted the hopelessness of their cause. The Boer negotiators tried for a time to achieve some form of sovereignty, but it became obvious that this was impossible. Nonetheless, the Treaty of Vereeniging was by no means an unconditional surrender. For instance, Afrikaans and English were given equal status in the new colonies.

Thus Britain acquired two more colonies, which were placed under Governor Milner. The Afrikaners, though defeated, were not united, and there was much bitterness between various factions; those who wanted to fight to the finish disagreed violently with the signing of the Treaty of Vereeniging, and Boers everywhere in the two ex-republics resented the failure of many of their Afrikaner friends in the Cape and Natal to come to their aid. In a way, this division was a useful tool for the British government in instituting the desired forms of government, but Milner set about this matter in a poor manner. Like Somerset at the Cape, Milner wanted to bring the English way of life to the Transvaal and the Free State, and while he did not prohibit the use of Afrikaans, he did create a constitution which prevented the Afrikaners from dominating the political scene. This and other matters helped to unite the Afrikaners once again.

In 1906, shortly after the departure of Milner, the Liberal party, which had disapproved of the Boer War, came into power in London, and responsible self-government was granted to the Transvaal and the Orange Free State. This did not mean that the republics were re-established, but, under British sovereignty, internal government was once again in the hands of the local people, as it was at the Cape and in Natal. This paved the way for eventual unification of the four areas.

Unification came earlier than might have been expected. In October, 1908, a National Convention held in Durban brought together the leaders of the Cape, Natal, and the former republics. The delegates acknowledged the need for some form of cooperation and removal of customs barriers, but the question to be solved concerned the nature of this cooperation. The two possibilities were a loose federation of virtually independent territories, on the one hand, and a close union, on the other, in which the four territories would have a central government. Significantly, the Cape had developed a local government with practically universal franchise, and there was lengthy discussion on the matter of the inclusion of non-whites in government. It was clear that the former republics would never agree to the inclusion of non-whites in the government, and it was agreed that the system of the Cape was to be abandoned and that governmental matters would be exclusively dealt with by whites. Doubtless those who gave in on this matter felt that the time would come when the Cape system would again find favor, but in this they have not up to the present proved correct. The choice of the capital city for the new Union was another problematic matter, and it was at this conference that the peculiar arrangement of today was created. Cape Town was made the legislative capital and Pretoria the administrative capital of South Africa, and the entire government travels back and forth each year.

In 1909, the British Parliament embodied the decisions of the National Convention in what is known today as the

South Africa Act. On May 31, 1910, South Africa's four colonies became the Union of South Africa, barely eight years after the last shots were fired in the bitter Boer War. Needless to say, there was opposition in South Africa to union, but many Boers worked for its success and survival. In the years to come, the Union was to pass through difficult times. Some of the difficulties were created by the South Africa Act itself: the elimination of the non-white voter was to have serious consequences. And perhaps union came too soon. There was general agreement about the need for economic cooperation in South Africa, but union forced the wedding of parties which had but recently fought bitterly and suffered severe losses. Union re-established the British rule which the trekkers had sought to escape, and it brought Englishman and Afrikaner face to face at a time when wounds might have healed better in prolonged separation.

The first Prime Minister of the Union of South Africa was a Boer, General Louis Botha. The British representative was the Governor-General, at this time Lord Gladstone. Botha as well as Smuts worked for real unification and real conciliation of Boer and Briton, and there were many people who wished to eradicate past grievances and to whom the success of the Union was the first objective. However, a war lost is more readily remembered than a war won, and almost immediately after union, Afrikaner nationalism began to assert itself. In the 51 years the Union existed, this nationalism played an increasingly important role, as will be seen subsequently. What is important here is the fact that in the political framework of Southern Africa, South Africa, made up of the four provinces of Natal, Transvaal, the Orange Free State, and the Cape, by 1910 had taken shape.

Across the Limpopo

While the sequence of events leading up to the Union of South Africa was in progress, matters were taking their individual course beyond the boundaries of the Transvaal. The northern and western boundary of the Transvaal is formed by the Limpopo River, which separates it from Bechuanaland in the west and Rhodesia in the north. Rhodes's Pioneer Column was establishing forts (the first at Fort Victoria) and raising the British flag over Southern Rhodesia. In 1891, the protectorate over Bechuanaland was established.

While the British South Africa Company, which was responsible for promoting trade and establishing government north of the Limpopo River, worked actively in Southern Rhodesia and in Northern Rhodesia to some extent, missionary societies penetrated Nyasaland, where the horrors of the slave trade were fresh in the minds of Africans and whites alike. In 1891, the slave trade was finally terminated and the tribal warring brought to a halt, and a protectorate was proclaimed over this territory. Nyasaland remained largely free of white settlement and exploitation. The chiefs kept their power, no major concessions were initially extracted from them, and the few whites on the scene were mainly missionaries and administrators. No major conflict occurred.

Northern Rhodesia, though like Southern Rhodesia administered by the British South Africa Company, escaped direct exploitation for the time being. Northern Rhodesia was not overpowered but was gained by the Company through negotiation and treaties. There was less actual conflict than there had been in stopping the tribal fighting and ending the slave trade in Nyasaland, and traditional rule was not seriously disturbed. To the north of the Zambezi River, land alienation was rare and was in fact discouraged on the principle that African interests came first in a protectorate.

In Southern Rhodesia, meanwhile, there were several causes for friction. Although the Company ruled the area, nothing prevented Boer farmers from settling in the territory save some restrictions on the lands they could occupy. As in the Transvaal, the arrival of the white man caused trouble with the Africans, and Southern Rhodesia was more densely populated by the Matabele and Mashona than the greater part of the Transvaal had been when the trekkers first arrived there. The Matabele, having been involved in the wave of Zulu aggressions, were a raiding people, who existed on the basis of attack and plunder. They raided the Mashona frequently, and it was among the Mashona that many white men came to settle. These raids by Matabele on Mashona endangered the property of whites, affected their servants, and made for generally unsettled conditions. When, in 1893, the Matabele attacked an outpost of the Company also, war broke out. There were less than 700 white men capable of carrying arms in Southern Rhodesia in 1893, but the Matabele were quickly defeated. King Lobengula's capital, in the south, was occupied, and Lobengula himself lost his life while in flight. The city of Bulawayo was founded near this spot the year after the defeat of the Matabele.

Three years later, the Jameson Raid (December 31, 1895, and into 1896) brought trouble to Rhodesia as well as the Transvaal. The bulk of the police force of Southern Rhodesia went with Jameson's army into the South African Republic, and this left Southern Rhodesia practically defenseless. Consequently, in 1896, the Matabele, now allied with the Mashona, rose once again, in the belief that the defeat of Jameson's troops was a sign that they could recapture Rhodesia. The whites had been training African "loyal" police, and many of these were still in Rhodesia. However, many of the African police joined the uprising, and even without them the Matabele were still strong despite the defeat of 1893. Taxes, cattle raids, and labor conscription were all sources of grievance, and Africans of many tribes collaborated this time and did not leave the fighting to the Matabele alone. A costly war resulted, and it is interesting to note that the same British who were so adept at intervening to stop the fighting between Boer and African in the republics in the interests of peace were so hated in Southern Rhodesia. The fighting

lasted somewhat less than two years, for the settlers,
though greatly outnumbered, had the advantage of supply
from the Cape and Britain. Supplies had to come 500
miles by ox-drawn wagon from the nearest railhead, which
at that time was at Vryburg, and the Matabele harassed
the wagon trains, but eventually supplies made the difference between victory and defeat for the white men. In
1897, after their lands had been destroyed, kraals burned,
cattle driven away, and many leaders killed, the Matabele
surrendered. The Mashona, somewhat farther away from
the main force of the white settlers, fought on, but they
too were finally subdued in the same year. The defeat of the
Matabele and Mashona in this war was partly the result of
physical conditions which have greatly aided the white settler throughout his occupancy of the subcontinent. Unlike
Kenya and other African countries, the plateau of Southern
Africa is open grassland with savanna along the slopes;
hiding places are relatively few, and dense forests rare.
Guerilla warfare is not easily carried on under such circumstances, and insurgents are easily located and flushed
out.

Since 1895, by Imperial sanction, Zambezia had been
known as Rhodesia, in recognition of the contributions
made by Rhodes to the development and spread of British
influence in this region. During the years immediately after
the final war with the Matabele, economic progress was
considerable, and among the important events was the development of a railroad system which connected Southern
Rhodesia with the coasts. The connection of Bulawayo
with Kimberley via Bechuanaland was established in
1897, and Salisbury was linked with the Portuguese port of
Beira by rail in 1899. Two years after Rhodes's death in
1902, the Zambezi was bridged at Victoria Falls. In 1901,
meanwhile, Salisbury and Bulawayo had been connected by
rail, and the two isolated regions of Matabeleland in the
south and Mashonaland in the north were linked with each
other as well as with the outside world.

The Boer War temporarily slowed development in

Rhodesia. Over 20 per cent of the white male adult population was involved in the fighting. Farms and mines were abandoned or at least temporarily closed down, and the rail connection with the Cape was interrupted. The influx of settlers was halted, though no major African uprising took place, and the crisis did not reach the proportions of 1896. Since 1902, however, progress has been continuous. Although hopes of another Witwatersrand on the Rhodesian plateau did not materialize, gold, asbestos, chromium, coal, and other minerals were discovered here and there, and with the development of the cities, secondary industry developed, and agriculture was stimulated. Settlement expanded at an increasing rate, particularly after the discovery of preventives against malaria and blackwater fever, and forts began to change to towns. Bulawayo became the most important center of the Rhodesias, and in Salisbury, where the organization of administration was centered, things improved also.

The similarity in development between interior South Africa, on the one hand, and Southern Rhodesia, on the other, is obvious, but the difference in timing is important. Africans were being subdued, minerals discovered, and towns founded in Rhodesia after the diamond industry in South Africa was already 30 years old and the city of Johannesburg had existed for a decade. In some ways, Southern Rhodesia has never caught up; the white population of the territory numbered about 225,000 in 1961, less than that of the Orange Free State. Salisbury is not comparable in size to Johannesburg, Durban, Cape Town, or a number of other cities in South Africa. Mining, urbanization, and industrialization commenced earlier and proceeded faster in South Africa. In Southern Rhodesia as in South Africa the native peoples were conquered and have been subjected to white rule, but Southern Rhodesia touches on Northern Rhodesia and Nyasaland, where white domination fades perceptibly.

By 1910, the year of union in South Africa, Southern Rhodesia had taken on the appearance of permanence and no longer seemed a frontier. Land settlement was encouraged, particularly after 1907, with favorable results, and the white population by 1910 may have been as high as 10,-000. Voices were calling for increased independence for the territory and a greater share in local government.

In Northern Rhodesia, meanwhile, the Copperbelt had been discovered, and in 1904 a start was made with a railroad to link Livingstone to the Congo. Small exports of copper, zinc, and lead were beginning to trickle out of this region, and along the railway, farmers began to settle. Although the government objected to this exploitation of a protectorate, the Company favored this development for economic reasons, and a small nucleus of white settlement began to develop in the heart of the Africa that had been acquired not by force but by negotiation. Some land alienation took place, and from this period on, northwestern Northern Rhodesia took a somewhat different line of development from eastern Northern Rhodesia, where mining was negligible and farming was not stimulated by the development of markets. One of the achievements of this period in the more prosperous area was the diversification of agricultural production. Coffee, and later tea, were introduced into Nyasaland. Tobacco was introduced in both Rhodesias, and cotton was also grown. Farmers were attracted by the Company, which aided them with loans, and labor, if unskilled, was cheap.

As the economy of Central Africa developed and the population grew, the demand for political power by the local whites became stronger. Salisbury and Bulawayo were the first towns to become municipalities, controlled by the ratepayers, in 1897. The following year, a Legislative Council was established in Southern Rhodesia. The electorate was almost entirely white, but Africans who could satisfy the requirements could be placed on the voters' rolls. In the Legislative Council, which was made up at first of a majority of appointed members from the Company together with some elected members, the balance changed as the years progressed. After Rhodes's death the number of appointed Company members was the same as that of the elected members, and subsequently the elected members attained the majority. London rarely intervened in the actions of the Council, and the influence of the Crown was not much felt.

As Southern Rhodesia sped toward self-government, the political situation in Northern Rhodesia and Nyasaland changed little. There was no great influx of whites here, and many of those who did come were missionaries and administrators, not miners and farmers. The white minority did not attain great economic strength and did not wield much political power. In times to come, the differences between the Rhodesias north and south of the Zambezi were to be felt throughout Southern Africa.

| The Desert Coast

While the chief drama of Southern Africa was played out on the interior plateau, events were taking place, hardly noticed, to the west of this main stage of activity. The western regions of the subcontinent were known from the early days to be desert or steppe lands, undesirable compared to those of the eastern plateau. The coast north of Cape Town, and especially north of the point where the Orange River reaches the sea, was deserted and had no natural harbors comparable to those at the Cape. Many a ship had been destroyed on the reefs off this coast. Although Diego Cao, the Portuguese navigator and explorer, had voyaged along the coast in 1484 from the mouth of the Kunene River in the north to Cape Cross in the south, there was nothing to attract overland exploration. It was not until 1760 that the area was first penetrated by a white man. His name was Coetse, and he brought back details of the dryness of the land and the nature of the people he encountered. In 1792, soon after the first settler had come to the territory, Van Reenen made the initial contact with the people whom Coetse had heard about but never seen: the Herero of the north. In the same year, the Dutch, then still in power at the Cape, took posession of the best inlet along the entire coast, Walvis Bay. It had been reported by the few travelers and settlers in the region that the Herero possessed great herds of cattle, and occupation of the inlet at Walvis Bay seemed a profitable move. Walvis Bay did not change hands when the British took over the Cape, and activity in South West Africa, as the area came to be called, was at a minimum. Developments at the Cape and its frontiers were all-important at that time.

By 1802, the London Missionary Society, which played such a significant role in events at the Cape, was showing interest in South West Africa. The Society employed not only Englishmen and Scots but also people of German and Dutch extraction. The first missionaries to be sent to South West Africa settled in Namaqualand, and they were the brothers Albrecht, of German descent. Others soon to follow were Schmelen and Moffat, and the number of German-speaking people in South West Africa began to increase.

At the time of arrival of these missionaries, South West Africa was feeling the pressure of the Bantu invasion. The Hottentots of the area had not been affected by years of contact with the whites, whose settlements up to this time all lay south of the Orange River, and the Hottentots had driven the Bushmen into the Kalahari wastes, but from the north came a threat to their domination. This challenge came from the Herero about whom Coetse had vaguely heard and who in the half century after Coetse's travels had grown in numbers and strength. The Herero comprised one of the Bantu groups moving southward. Other Bantu peoples who had reached South West Africa were the Damara and the numerous Ovambo and they all coveted the rich pastures of the northern part of South West Africa. The Hottentots had a leader as formidable as Chaka of the Zulu and Lobengula of the Matabele. He was Jonker Afrikaner, a feared and powerful chief, and it was he who organized the Hottentots into successful armies which kept the Herero at bay. Afrikaner died in

1860, and with his death Hottentot power began to fade. The 1860's in South West Africa were times of war and terror. The Herero at length defeated the Hottentots, and the constant lack of safety reduced the white population, missionary and otherwise, to a few individuals. In 1868 those left in the territory requested British protection, and petitions were signed by most adult whites and sent to the Cape. Frere, then the Governor at the Cape, did his best to convince the British government of the need for annexation, but the sole result of his efforts was the annexation by Britain of Walvis Bay and some 400 square miles around it. This occurred in 1878, when British interest in Southern Africa was focused elsewhere.

Serious conflict between the African peoples broke out again in 1880. Since the first appeals of the settlers in 1868, the German government had shown some interest in the events in the territory and had corresponded with Britain on the matter of protection of the white settlers. London never did commit itself thoroughly on the matter, and there was some doubt concerning the extent of the area for which Britain felt herself responsible. In 1883, an agent acting on behalf of Lüderitz, a German merchant, obtained concessions to some 200 square miles around Angra Pequena, South West Africa's second navigable bay and landing place. Angra Pequena came to be called Lüderitz, as it is today, and formal German acquisition of the territory had begun. Until this time, South West Africa had been viewed as a commercial hinterland by the Cape Colony government, which did not believe that anyone but Britain would lay claim to it. Thus, when Bismarck announced in 1884 that the possessions claimed by Lüderitz were under the protection of the German Empire, there was considerable consternation in Cape Town and London. Governmental machinery was set in motion for the annexation of the South West African coast from the Orange River to the area claimed by the Portuguese (approximately the Kunene River), but before any action could be taken, a German warship appeared at Angra Pequena

(Lüderitz Bay) to hoist the German flag. A German protectorate was proclaimed.

The Herero, soon seeing the danger of German occupation, desperately attempted to obtain British protection. Their efforts were to no avail, however, and when Britain failed to extend her possessions from Walvis Bay into the land of the Herero east and north, Germany took these lands in 1885. The old headquarters of Jonker Afrikaner became the seat of the capital, Windhoek, after some skirmishes with the Africans. African power was not entirely broken, however, and in 1893 a series of long and bitter wars broke out between the Africans and the Germans. The British referred to German actions in South West Africa as "colonisation by the Mauser." It is important to note that British and Dutch farms were not attacked but that German establishments of all kinds were destroyed by the African guerilla rebels. Angered by their failure to end the fighting and subdue the Africans, the Germans embarked upon a series of actions which have left their indelible mark on the territory's people. In all African history, no extermination campaign wrought more horror and caused more lasting resentment than the German operations against the Herero. Thousands of Herero were shot, whether they were armed and opposing the Germans or not. Many more were herded into the desert wastes and left to die of thirst. Estimates of the loss of life range from 30,000 to 50,000, and historians tell of acres of territory covered with the bleaching bones of victims of the campaign. In 1905, the proclamation under which this killing was taking place was repealed, and amnesty was granted to the Herero, who responded by laying down their arms, if any, and moving into reserve areas. The Hottentots carried on the war until 1907. Nearly 20,000 German soldiers had been in the field when the campaign was at its height—a force equal to the entire white population of whole provinces of Southern Africa.

By 1910, South West Africa was changing for the better. Diamonds had been discovered near Lüderitz Bay, and the value of the area was being recognized. Changes in administration after 12 years of war brought some progress. Initial steps toward government (white only) were taken. An Advisory Board of 40 members (20 appointed by the Governor and 20 elected) sat in Windhoek to advise the Governor, who was himself appointed by the Kaiser. This was an arrangement reminiscent of the Cape a hundred years previous. South West Africa, however, was not to experience, under German, rule, the evolution of responsible representative government.

South West Africa extends from the Orange to the Zambezi River. While the Orange simply forms the southern boundary, the territory reaches to the Zambezi by means of a long, eastward proruption, one of the anomalies of the political map of Southern Africa. The Caprivi Strip, as the proruption has come to be called, was the result of a desire on the part of the Germans to gain access to the "route to the east"—for which the Zambezi is less useful than the Germans supposed. When the Germans in 1890 signed an agreement by which it was "understood that under this arrangement Germany shall have free access from her Protectorate to the Zambezi by a strip of territory which shall at no point be less than 20 English miles in width," they bequeathed a political liability to South Africa.

| *The Basuto and the Swazi*

A glance at the political map of Southern Africa will reveal the existence of two small units, Basutoland and Swaziland. Basutoland lies between Natal, the Orange Free State, and Cape Province and is completely surrounded by South African territory. Swaziland, while almost enclosed by the Transvaal, has a short but crucially important border with Moçambique.

Basutoland and Swaziland are two of the three High Commission Territories in Southern Africa (Bechuanaland is the third), and their origins are closely tied to events involving the Zulu wars and the invasion of the interior by the Boer trekkers. The Basuto, a nation developed from those fleeing before Zulu might, today still occupy the mountain stronghold where Moshesh united them and fortified them against attack. Basutoland lies to the southeast of Zululand, atop the Drakensberg and stretching west into the lower plateaus of the interior. After first surviving the Zulu onslaught, Moshesh and his people were faced with the new challenge of the white trekkers. In 1843, the Basuto chief signed a treaty with the Cape Governor, placing himself and his people under the British Crown. The Basuto realm as defined at this time roughly equaled the area of Basutoland today. Soon afterward, there was trouble between the Boers and the British in the region between the Orange and the Vaal, and the British were victorious. After the British had defeated the Boers, Major Warden, the resident commissioner, attempted to demarcate the boundary between what was to be the Orange River Sovereignty and Basutoland. His efforts were not appreciated by several minor groups of Basuto who found themselves cut off from Basutoland and from their chief, Moshesh. Friction arose, and the British sent a small contingent of men to enforce the rule. In the war that followed, the Basuto demonstrated that after Chaka, Moshesh was the most powerful chief ever to partake of the South African drama. The British troops were defeated (even though some historians do not realize it), and the reinforcements sent from the Cape—over 2,500 men, a very strong army for those days —were also in a very poor position when the fighting ended. The termination of hostilities came when Moshesh, in a relatively strong position though no doubt increasingly vulnerable with time, wrote to the British authorities that the British had shown their tremendous strength and should now also show their leniency by calling a halt to the war. This much-discussed document requesting clemency from a position of strength led to the recognition of Basuto independence at the 1854 Convention between the Orange Free State and Britain at Bloemfontein, where the independence of the Boer republic was also recognized.

It was not long before the Boers of the Orange Free State and the Basuto fought over the boundary between their lands. The Basuto, who had carried on successful guerilla war from the safety of the high plateaus of the Drakensberg, desperately needed the western part of their country for pasture lands, which are rare along the barren slopes of the high mountains. The frontier between the Basuto and the Boers was one where grasses were rich and the land desirable, and in 1856 war broke out over it. The fighting was most intense till 1858, when the Basuto, in a weak position this time, resorted to sporadic raids and attacks. In 1861 the Basuto petitioned for British protection in the face of the increasingly successful republicans, but the response was negative. Meanwhile, the war continued, and in 1865 it became more intense again. This time, when asked by the Basuto for protection, the Cape Governor sent a mediator who attempted to settle Boer-Basuto differences. In this, there was no success, and in 1866 the Boers and the Basuto signed a treaty which was little more than an unconditional surrender on the part of the Basuto, who came under the rule of Bloemfontein.

This treaty did not stay in effect very long, for in the very next year war was renewed. Finally, when Moshesh again petitioned for British protection, in 1868, the Governor of the Cape issued a proclamation whereby Basutoland was placed under British guardianship. Maseru was made the capital, and a police camp was established there. Moshesh had finally achieved what had been his major aim since his power began to wane: his country was taken over by Britain. His perseverance was well rewarded, for when union came about in South Africa in 1910, Basutoland remained separate, and even today it is still under British care a black man's country in turbulent South Africa. The old chief died at the age of about 90 in 1870, but in the Basutoland of today his works have their effect and his spirit survives in a strong national pride.

The fact that Britain had taken over Basutoland did not mean that all the Basuto's troubles were now over. In

1871, the territory was annexed to Cape Colony, contrary to what had been Moshesh's wishes. Initially, the stimulation of all kinds of activity as a result of the diamond discoveries also affected Basutoland, and there was a period of considerable economic progress. Demand for grain and labor was sharp, and many Basuto came back prosperous from the diamond fields. The soil erosion which has since become perhaps Basutoland's greatest domestic problem began in this period as agriculture developed, but there was no land shortage in the 1870's. Another effect of the diamond fields and the comparative wealth many of the Basuto (and Africans elsewhere in the subcontinent) was that many firearms were bought by tribesmen. Before long, the Basuto, from a defeated and destitute people, became the best armed nation in the south of Africa. Their newfound power was expressed by a desire for direct representation in the government at the Cape. On grounds which in retrospect worked to the advantage of Basutoland, the request was denied.

The last of the Kaffir Wars took place along the frontier of the eastern Cape in the early 1870's, and this conflict had some minor reverberations in Basutoland when some of the Cape and Natal chiefs sought refuge there. However, they were followed and arrested with very little fighting. Unfortunately, the war did have a sequel.

The defeat of the Africans in East Griqualand and the death of the Griqua chief, Kok, focused Basuto attention upon this area, and there was reason to believe that the Basuto chief, Nehemiah, was plotting to extend the territory of Basutoland. Nehemiah was temporarily jailed in 1876, and though he was found innocent and freed in 1877, the embittered Basuto began to destroy white property and threaten white residents in their country. The period until 1880 was marked by increasing tension. Efforts to disarm the Basuto went awry, and after a number of introductory skirmishes a major conflict broke out in 1880 between the Basuto and the British. The actual heavy fighting lasted less than a year, and negotiations were opened in 1881,

after the Basuto had suffered some heavy defeats but not
before the white troops, also, had suffered comparatively
heavy losses. By the end of 1881 it was clear that the Basuto were not going to win the war against white over-
lordship. However, the negotiations broke down repeatedly, and the Basuto were unmanageable; in short, Cape
Colony was unable to handle the territory. In 1883 the ill-
fated annexation of Basutoland to Cape Colony came to an
end, and in 1884 the London government took over responsibilities for the Basuto. This meant that the country was
semi-independent, the authority of the chiefs was unim-
paired, and the Basuto had, in effect, won their battle
against domination from the Cape.

The ensuing years saw progress in Basutoland once
again. The Boer War created a great demand for the Basuto's renowned ponies, and grain, meat, and other supplies
were required by the British Army. With the end of the
war, soil erosion, overgrazing, and land shortages were beginning to make their appearance. However, the development of mines in South Africa drew numbers of Basuto to
the new sources of wealth, as it does today. Improvements
were made in medical facilities, education, and communications. When the surrounding territories in 1910 became the
Union of South Africa, Basutoland remained the separate
entity it is today.

The history of Swaziland, slightly over half the size of
Basutoland, is a less violent one, although it includes some
great battles with the Zulu, the mutual enemy of the Basuto
and the Swazi. Major armed conflict with whites over
Swazi soil did not take place, although a sequence of careless concessions made by Swazi chiefs robbed the Swazi of
much of their land. This land, which lies along the escarpment in the southeastern Transvaal, was probably first settled about 1750 by Chief Ngwane III and his people. From
1815, when the Zulu were gaining in power immediately to
the south of Swaziland, Chief Sobhuza I ruled the Swazi
until 1839, and under his leadership the Swazi became a
force to be reckoned with in Southern Africa—strong

enough to withstand the onslaughts of the Zulu, who were attacking surrounding peoples in all directions. Both Chaka and Dingaan made attempts to subdue the Swazi, and although the latter were not sufficiently numerous and strong to defeat the Zulu in open combat, they successfully resisted Zulu occupation of their territory. In the meantime, the Zulu were defeating tribes to the west and northwest of Swaziland, and when Mswazi II succeeded Sobhuza I in 1839, Swaziland entered into a difficult period in her battle for survival. One of the changes Mswazi II brought about was the moving of his capital farther north and thus farther from the Zulu might, thereby decreasing the possibility of its destruction. In those days, the territory occupied by the Swazi was larger than the Swaziland of today.

Mswazi II took power in Swaziland just as the battle of Blood River ended in the defeat of the Zulu, and one of the effects of the Zulu defeats in southern Natal was increased activity in the north. Dingaan's power was not completely broken, and he was strongest in northern Zululand, immediately south of Swaziland. When Panda, his half-brother, joined the Boers in Natal, Dingaan was again defeated, and he was actually killed by Swazi warriors when he attempted to escape across the Lebombo Range, along the crest of which runs the present eastern boundary of Swaziland. Dingaan's death and increased pressure from the south on the Zulu caused an increase in the frequency of Zulu raids and invasions into Swaziland, and Mswazi II took the action so many African peoples in the subcontinent took in the 19th century: he appealed to Shepstone in Natal for British protection. Shepstone was unable to extend this protection, but he did demand that Panda and the Zulu cease their attacks on the Swazi. He was remarkably successful in his negotiations, and the series of hostilities between Zulu and Swazi came to an end.

A new episode commenced very shortly afterward when, still during the reign of Mswazi II, the first whites appeared in the southeastern Transvaal with the intent to settle permanently. Events in Natal contributed to this desire on the part of the trekkers, and Swaziland contains some of Southern Africa's best pastures. Until about 1845 the whites who had visited Swaziland had been mainly hunters and traders, but in that year land alienation began. In 1846, Mswazi II signed a treaty with Potgieter, one of the leaders of the trekkers, in which he ceded to the Lydenburg Republic (one of the four units which were to form the South African Republic) all rights to all land north of the Crocodile River, in exchange for 100 head of cattle. This concession, to be measured in hundreds of square miles, perhaps a thousand, was the first of a series. Between 1845 and 1900, these concessions caused problems which even today have not been settled in Swaziland. The major culprit in this respect was Chief Dlamini IV, who came to power seven years after the death of Mswazi II and who ceded rights of various kinds for the remainder of his land, in return for such items as greyhounds, gin, money, and cattle. Not only were land grants made, but the chief pledged exemption from all taxes on whatever establishments might be erected. In a final concession, all land not previously allocated was ceded. Initially, the whites were primarily interested in the fine grazing lands in the hills of Swaziland, but in the early 1880's gold was discovered in the Barberton Mountain Land and soon mineral rights were among the concessions the chief distributed for trivial returns.

In 1881, at a convention between the South African Republic and Great Britain, both sides recognized the independence of the Swazi within certain boundaries. In a superseding 1884 convention, these boundaries were re-delimited and described in full. A map of the South African Republic of the late 1880's resembles almost completely one of the Transvaal of today as far as the boundaries with Swaziland are concerned. The boundary with the Portuguese was surveyed in 1884 and later years, and it was finally demarcated in 1907. The conventions of 1881 and 1884 were the first in a lengthy series, all with the object of settling further problems which kept arising with regard to the Swazi. In terms of the convention of 1894, the powers

of administration and jurisdiction in Swaziland were dele-
gated to the South African Republic. This occurred after
the Swazi, in 1887, had appealed in vain for British protection once again: the country was being overrun by fortune-
seekers. For a short time (1889–91) there was a sort of
independence for the whites in the territory, and in the
southwest a small Afrikaner "colony" formed and appealed
successfully to the 1890 convention for inclusion in the
South African Republic.

After some years of administration by the South African
Republic, the Boer War terminated the arrangement, and
the Transvaal was annexed by Britain. Swaziland thus
came under British administration by default and was ruled
by the new Governor of the Transvaal after the Treaty of
Vereeniging had been drawn up. In 1906, when South
Africa had returned to a more normal state, an Order in
Council transferred all powers over the territory, which
were in the hands of the Governor of the Transvaal, to a
High Commissioner for South Africa. When the National
Convention was held in Durban to deliberate union in
South Africa, the Swazi requested that they be excluded
from any such arrangement. Their request was granted,
and their country retains today the status of protectorate.

Their political future somewhat more secure at the end
of the first decade of the 20th century, the Swazi were still
faced with the legacy of concessions which affected practically all their living space. A commission had been appointed in 1908 to deal with this matter, and its first move
was to deduct one-third of the land granted in each individual concession. This land was to be returned to Swazi
hands, but all Africans would have to move from the remaining white-owned land within a period of five years
after 1909. They could, after 1914, remain on white-owned
land only by arrangement with the concessionaire or his
successor. Most Africans managed to come to satisfactory
terms with the land-owners, and no major migration off the
white lands appears to have taken place. Presently, the
Swazi are buying back land that still remains in the hands

of whites, and the African-white ratio of land ownership is becoming more favorable to the former.

As a result of the scattering of adjacent white and African land parcels, a good deal of interracial cooperation was necessary in Swaziland, so that, although the concessions have been a real liability to the Swazi, they have also been an asset by fostering good will between neighbors black and white.

| Portugal in Southern Africa

From the Congo to the Kunene River and from the Rovuma River to beyond Delagoa Bay, Portugal possesses the coasts of Africa. The two and a half thousand miles of coastline of Angola and Moçambique are by no means rich. Natural harbors are few. The coastal plain is narrow and desert-like in Angola, hot and humid in Moçambique. The interior, where owned by Portugal, is not particularly prosperous. Nevertheless, these "Provinces" are among Portugal's greatest assets and among its oldest possessions. They give Portugal a share in the control of three of Africa's greatest rivers, the Congo, the Zambezi, and the Limpopo. They include the natural outlet for Africa's richest and most exploited mineral regions, the Witwatersrand and the Katanga Copperbelt. They and the Central African Federation, linked to them in economic cooperation, constitute today's buffer zone between black Africa and the "white south."

Portugal laid claims to its current African domain long ago. Angola was the first European colony in central Africa. Its coastline was explored by Diego Cao as early as 1483, and Vasco da Gama put ashore in Moçambique in 1498. The first settlement to be established and fortified was Sofala, located on the coast a short distance south of the present port of Beira. By 1505, this place was a stronghold of Portugal, and in the next decades it grew to prominence as the hinterland was explored and exploited. Slaves and gold passed through on their way to Lisbon and elsewhere, and Sofala became a port of call for ships on their way to the Indies. Farther up the east coast, opposite Lake Nyasa, is the town of Moçambique. It was founded in 1508 and fortified between the years 1508 and 1511, and it has never been out of Portuguese hands. Moçambique attained importance in the 16th century as a slave port where Nyasa Africans were herded aboard ships bound for other continents.

While there was considerable commercial activity along the coast of Portuguese East Africa, Angola was relatively dormant after the early explorations of Cao. The major activity was religious, and many missionaries made their way to Portuguese West Africa to convert the heathens to Christianity. In this they appear to have had considerable success, and in time many of their converts got a first-hand sample of Christianity in action when they were taken to the Americas as slaves. In 1549 the first white women were shipped to Angola, and permanent white settlement had begun. In 1575, Governor Diaz founded Luanda, today still the first city of the territory.

In Portuguese East, meanwhile, further explorations were made. In 1502, Antonio de Campos, one of Vasco da Gama's captains, discovered Delagoa Bay, one of the world's best natural harbors. No further exploration was done at Delagoa Bay until Lourenço Marques, a merchant trader, explored the inner reaches of the estuaries leading into it. Between 1544 and 1550 the value of this bay was realized, and in recognition for his work in this area, King John III ordered that the bay be renamed Baia de Lourenço Marques. The Portuguese did not have the success they had at Sofala and Moçambique, however, when it came to establishing trading stations at Lourenço Marques. Repeatedly, the Africans wiped out their settlements, and white occupation at Lourenço Marques was intermittent. Portuguese attention during the second half of the 16th century was concentrated on the northern part of their eastern domain, since they came into contact with the Arabs who still dominated this portion of the coast. Sev-

eral conflicts took place during the 1580's, and one of the
main scenes of action was the port of Mombasa, which
changed hands several times and was the cause of some
major battles.

Partly through contact with the Arabs, the Portuguese
began to take an interest in the interior, which was thought
to be very rich. Gold was the main object of the traders,
and as early as 1531 the Portuguese had advanced up the
Zambezi River to Tete, where they established a fort. As
early as 1620, about two and a quarter centuries before
Livingstone, Bocarro is reputed to have crossed Lake
Nyasa. Possibly Portuguese and Arabs first reached the
Zimbabwe ruins in Rhodesia and wrought the havoc there
which has been deplored so strongly by anthropologists and
archaeologists. Settlements, however, remained confined to
the coast, and in the 16th century there was no expansion
into the interior such as the Cape of Good Hope was to
experience.

In Angola, there was no contact with any peoples but the
Africans, and the interior did not have the reputation of
great wealth. Angola was not so favorably located with
respect to the route to the Indies, and it is not surprising
that no major effort at penetration of the interior was
made until 1606, when De Aragao attempted to reach the
Zambezi headwaters from the west coast. Angola was
being used in those days as a place of exile for criminals,
the first group of whom arrived in 1593. Benguela was
founded in 1614, and during the first quarter of the 17th
century there was some immigration from Portugal despite
repeated revolts by the Africans. This period also saw the
decline of Portuguese power in Europe and the rise of
Holland, and the Dutch took possession of Luanda and
Benguela in 1640. In 1648, the Dutch were ousted from
Angola, but during the next 200 years Angola was neglected and served almost solely as a slave reservoir, for
which it had a "good" reputation. Although the territory
already occupied remained in Portuguese hands and the
interior was not claimed until the treaties with the British

and other interested parties were drawn up in the 19th century, little use was made of the opportunities for settlement. Some small parties settled on the Benguela plateau (the first in 1685), but the Portuguese in general showed as little inclination to develop Angola as the Dutch were indifferent to granting the Cape colonial status. The reasons were the same: if a territory was not rich enough to warrant occupation while exploitation was in progress, it became a liability, not an asset.

The general decline of Portugal was evident in Moçambique as well as Angola. After 1700, Mombasa having fallen to the Arabs following a 33-month siege, Portuguese establishments in the east decayed, trade collapsed, the Arabs threatened, the Bantu in the interior attacked, and missions were abandoned. What remained was the slave trade, which helped to revive Moçambique during the 1800's. For permanent settlement, the interior of Moçambique seemed even less desirable than that of Angola, and with the exception of small areas immediately surrounding the port towns and a few isolated parts in the north, white settlement did not expand. The British and Dutch in turn displayed an interest in the coast and even established stations at Lourenço Marques, but the general unattractiveness of the region led to their abandonment. In 1781 the Portuguese built a fortress on the site of the present city, but it was completely destroyed by Africans in 1796.

For both Angola and Moçambique, the mid-1800's marked the beginning of renewed Portuguese interest and activity. In Angola at this time began the real period of colonization which has not yet ended. Farms were established in the interior, the influx of whites increased in numbers, towns were enlarged, trade revived, and communications improved. It is noteworthy that this general improvement occurred after slavery had been abolished, although the Portuguese did not always adhere to abolition and hindered efforts on the part of the British to terminate slavery in, among other places, Nyasaland. In Moçambique, the old site of Lourenço Marques was the scene of renewed colonial activity. The Africans revolted but were defeated in the south in 1868, and the Portuguese government, aware of the growing importance of the interior, sent a commission to Lourenço Marques in 1876. Although the bay constituted a magnificent natural harbor, the land upon which any town could be established was interrupted by malarial swamps, and these had to be drained. The commission found Lourenço Marques to be a derelict assemblage of huts and stockades, but the place was declared a village, and in 1883 the railroad to the interior was begun. The Portuguese government had come to realize, first, that it possessed the natural outlets for an interior plateau that was constantly growing in importance and, second, that the first transport and communications system to reach this interior would reap great profits. The railroad from Lourenço Marques to Johannesburg was completed in 1895, and in that same year the Portuguese defeated the Africans so decisively that the victors, for the first time, effectively controlled the area. Lourenço Marques, which had been declared a corporate town in 1887, was made the capital of Moçambique in 1907. The harbor was greatly improved, and it has become important to the Witwatersrand. Lourenço Marques thrived on trade and tourism and rapidly grew to prominence among southern Africa's cities.

The Portuguese were involved in a number of boundary disputes with other powers in Southern Africa. Portugal's claim to all land north of 26° south latitude (along the east coast) was disputed by Britain, which also had an interest in Delagoa Bay, but after a brief resort to arms President McMahon of France, acting as mediator, awarded Delagoa Bay to Portugal.

Germany, meanwhile, had taken possession of Tanganyika, and the boundary between the two territories of Moçambique and Tanganyika was determined by treaty with Germany as lying along the Rovuma River. In 1891 a treaty signed with Britain limited Portuguese mastery along the Zambezi to a point at Zumbo. The Portuguese had laid claim to the intervening area between Angola and Moçambique, and it was largely because of Rhodes, who rapidly acquired the Lobengula concessions, that this claim was not substantiated. In the same year, the lands of the Barotse and the Kafue were added by treaty to Northern Rhodesia. In 1877 and 1878, Pinto had traveled onto the Zambezi-Congo divide, which led to Portuguese claims to this region, and the area extending as far east as 24° east longitude was consequently allowed to the Portuguese.

By 1910, therefore, the boundaries of Angola and Moçambique were as they are today, and the modern phase of colonial development had begun. Further railroad connections were to be made between the ports Lobito and Lourenço Marques with the mining areas of the interior, greatly stimulating the growth of these places and providing much revenue for the government. Beira was already connected with the Rhodesian plateau and Lourenço Marques with Johannesburg. However, white settlement in Moçambique and Angola was—and in many ways is still—peripheral, with large towns along the coasts and settlements along some transportation lines in the interior. Huge tracts of both provinces are hardly touched by the invasion of the white man, and this was even more true in 1910, when the major influx of white settlers was yet to come. The pattern of events to come had been established, however, and stability and effective government existed. For all practical purposes, the boundaries established, with or without concern for African tribal units, were permanent and still exist today with very minor modifications.

| The New Era

The year 1910, the year of union for the Transvaal, the Orange Free State, Natal, and the Cape, heralded a new era in the history of Southern Africa. The major event, of course, was the genesis of the dominant Republic of South Africa, which towers above all other political units in economic strength, population, wealth and material progress.

The end of the first decade of the 20th century also saw the end of bickering over frontiers, the beginning of virtually uninterrupted peace—even if peace meant the subjugation of African peoples—and lawful, organized economic activity. Although German rule was soon to be terminated, the territory of South West Africa has retained its boundaries and name, and many would allege that even internal conditions there are not very different from what they might be had the Germans stayed. The three High Commission Territories remained separate from South Africa as they are to this day. Moçambique and Angola were beginning to take on the function and character they possess currently. In Southern Rhodesia, Northern Rhodesia, and Nyasaland, after less than 30 years of white settlement, the pattern of the future was clear even before 1910. The political framework had been molded, and the complexities of Southern Africa were there then as now. The colonial policies of Portugal, Britain, and Germany prevailed over large regions, some forms of traditional African authority had survived in other, smaller areas, and clearly headed toward independence was the Union, which, even though under the British Crown, was already the white-ruled African country without a European motherland. By 1910 the land-grab had ended, and the period of internal development began. The political status of Southern Africa's territories had been settled (with the exception of South West Africa), and the whites were firmly entrenched in command of government, education, religion, and business. Needless to say, differences in historical development were reflected in what happened in the various political entities after 1910. Whereas the Cape had seen hundreds of years of white settlement, and parts of the South African interior as well as Natal many decades, the Rhodesias had only recently witnessed the arrival of white settlers. The Germans and Africans in South West Africa had just fought a bloody war. While the Basuto had long fought the white man and retained his land, the Swazi did not fight a war against the white invaders, and he had lost virtually all his land. Nothing stirred Bechuanaland, which seems to have gone on for centuries without even emerging from its backwardness. Long-neglected Angola and Moçambique, in response to happenings elsewhere in Southern Africa, entered a new stage of activity after lying dormant for centuries. Ports were being improved, railroads built, and towns expanded.

In 1910, Southern Africa resembled the rest of subsaharan Africa to a very large extent; it had, as it were, caught up with events to the north. In 1961, Southern Africa resembled—what? Its old self, the Southern Africa of 1910, not the rest of subsaharan Africa in our time. Elsewhere—in Nigeria, Uganda, Somalia, Chad—the clock kept moving, and thus today the south is a relic island of history, where, especially in Angola and Moçambique, a glimpse of past centuries can be caught. While the south progressed economically, the socio-political clock stood still, and attempts were even made to move it backward. The Portuguese provinces and the Central African Federation are exposed today in their northern regions to the effects of the Wind of Change. Still shielded is the south, increasingly alone in a new kind of Africa.

FURTHER READING

BARNES, L. *The New Boer War*. Hogarth: London, 1932.

BRYANT, A. T. *Olden Times in Zululand and Natal*. London, 1929.

CLARK, J. D. *The Prehistory of Southern Africa*. Penguin Books: Middlesex, 1959.

CLOETE, S. *The Mask*. Pocket Books: New York, 1958.

DE KIEWIET, C. W. *A History of South Africa*. Oxford University Press: London, 1941.

EDWARDS, L. E. *The 1820 Settlers in South Africa*. London, 1934.

GROSS, F. *Rhodes of Africa*. Cassell: London, 1956.

HERTSLET, SIR E. *The Map of Africa by Treaty*. 3 vols. Harrison: London, 1909.

LIVINGSTONE, D. *Missionary Travels and Researches in South Africa*. Harper: New York, 1858.

MARQUARD, L. *The Story of South Africa*. Faber & Faber:
London, 1954.
MC MILLAN, W. M. *The Protectorates.* (*Cambridge History
of the British Empire,* Vol. VIII.) Cambridge University
Press, 1937.
RITTER, E. A. *Shaka Zulu; the Rise of the Zulu Empire.*
Longmans: London, 1955.
SHAPERA, I. *The Khoisan Peoples of South Africa*. London,
1930.
WALKER, E. A. *A History of South Africa*. London, 1928.
WALKER, E. A. *The Great Trek*. London, 1934.

South Africa: Land and People

The Republic of South Africa is a large country. Its 472,-500 square miles compare to those of Texas, Oklahoma, New Mexico, and part of Louisiana. It is more than twice the size of France, yet in subsaharan Africa it is not among the giants. The Congo is more than double the size of South Africa, and Angola, the Central African Federation, and Sudan also exceed it in area. Although large, the country does not have a high total of population. There are just under 16 million people in South Africa, whereas smaller Nigeria has nearly 40 million, and France 45 million.

South Africa is also a varied and beautiful country. Extending latitudinally from just over 22° south to nearly 35° at Cape Agulhas, the country partakes of tropical Africa while being lapped by antarctic waters on its southwestern shores. From sea level (and some of the world's most beautiful beaches) it rises to the interior plateau, which culminates in the Drakensberg at Thabana Ntlenyana (11,425 feet). A variety of climates, ranging from true mediterranean at the Cape to tropical in the north and from desert to humid subtropical in the interior and along the coasts respectively, have carved a complex mineral-rich geology into some of Africa's most magnificent scenery. On the interior plateau, elevation moderates the other-

wise excessive heat of the lower latitudes to create a
healthy and desirable environment, and pastures on the
plateau rank among the best in all Africa. Good soils exist,
which, aided by irrigation, will grow apples, grapes, sugar
cane, citrus fruits, and vegetables. Gold, diamonds, coal,
iron, and a host of other minerals are mined. Although the
indigenous forests are depleted, some magnificent stands
remain. No impenetrable tropical forests grow here, and
the country is less affected than its northern neighbors by
tropical diseases and disease-carrying insects.

The great non-political problem of South Africa is where
to find sufficient water. Although the country is climatically
varied and agriculturally rich, compared to other nations
in subsaharan Africa, much of it is desert or dry steppe.
The great Kalahari Desert, the largest of the world's
expanses of sand, extends into the western Transvaal,
the northern Cape, and the western Free State, and steppe
conditions extend beyond the limit of the sand. The Kalahari does not look like a desert over much of its area: it is
rather a sparsely grassed land, which suddenly comes to
life when the unreliable and brief rains fall. But what the
Kalahari is not in appearance, it is in fact. Sandy soils and
intermittently flowing streams combine with a great variability of precipitation to bring disaster to would-be cattle
ranchers and their poor herds. Overgrazing, poor veld
management, and consequent severe erosion have scarred
the South African countryside from the eastern Transvaal
to the western Cape. The Orange River, the country's
greatest stream, receives the Vaal as its tributary in the
interior and proceeds toward the Atlantic Ocean in a channel that is constantly becoming more muddy and clogged.
Its mouth is being closed by a sandbar, and the river
dwindles considerably in traversing the northern part of
the dry western Cape. The time will come when the Orange fails to reach the sea. There can be no doubt that
much of the aridity of South Africa is the result of human
(African as well as white) carelessness, and although to-

day efforts at conservation are being made, there are those who feel that in places the damage is beyond repair.

The plateau surface of South Africa, grassy and pleasant looking, especially in the wetter east, is called the *highveld*. The word "veld" refers to the grassland. At the foot of the escarpment which leads down from the upper plateau surface is the *lowveld,* generally below 1,800 feet. The slopes of the escarpment, which in several areas such as Natal and Swaziland is steplike, are referred to as the *middleveld*. The word "veld" recurs in naming such parts of the landscape as *bushveld,* which is grassland interspersed with trees, characteristic of much of the lower middleveld, and *bankeveld,* which is the typical cuestaform topography found particularly well displayed in the vicinity of Pretoria. South Africa resembles, in a very general way, a soup plate which has been turned upside down: a narrow rim of lowlands, a steep scarp, and an extensive upper surface which, however, is depressed in some places, like the Kalahari basin, and elevated in others, such as the Basutoland plateau. Johannesburg lies on this plateau at about 5,500 feet, while Pretoria, just 30 miles to the north, is hundreds of feet lower. Although the highveld is the source region for several of Southern Africa's great rivers, it lies remarkably flat and is interrupted only by an occasional granite mound or some tilted sedimentary beds. In general, slopes are gentle, and the horizon is wide. Views of this kind are particularly characteristic along the road from the eastern Transvaal to Pretoria. In winter (which falls in June, July, and August in this part of the world), the veld turns gray, the skies are clear, the air is crisp and may even be cold. The atmosphere is dry, and so are many wells. Grasses lose their nutrient value. Grass fires are common. The whole environment takes on a quality that is not easily defined. These are the difficult months for the cattle farmer, and droughts are long and unbroken. With summer arrive the great thunderstorms for which the highveld is famous, and rain is concentrated in this

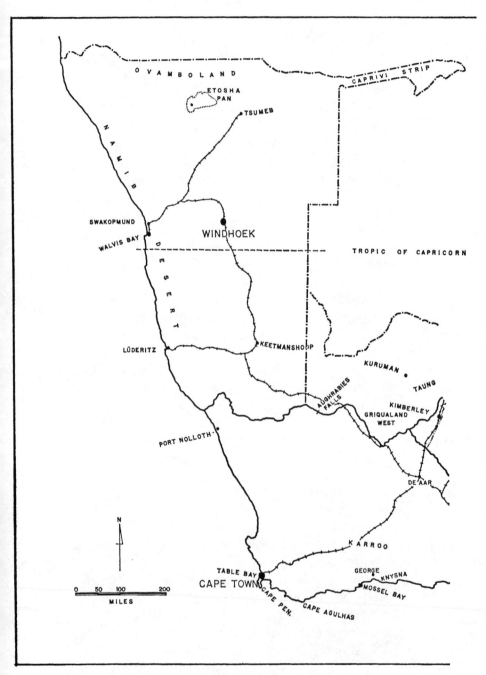

MAP 5 *South Africa and South West Africa Location Map.*

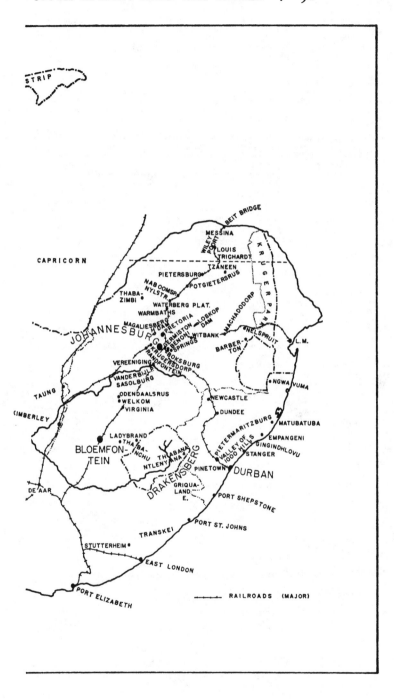

STRIP

BEIT BRIDGE

MESSINA
LOUIS TRICHARDT
CAPRICORN
TZANEEN
PIETERSBURG
POTGIETERSRUS
NABOOMSP.
NYLSTR.
THABA-
ZIMBI
WATERBERG PLAT.
WARMBATHS
MACHADODORP
MAGALIESBERG
LOSKOP DAM
JOHANNESBURG
PRETORIA
GERMISTON
WITBANK
NELSPRUIT
L.M.
BENONI
SPRINGS
BARBER-
TON
BOKSBURG
KRUGERSDORP
RANDFONTEIN
VEREENIGING
VANDERBIJL
NGWA VUMA
SASOLBURG
TAUNG
ODENDAALSRUS
WELKOM
NEWCASTLE
VIRGINIA
KIMBERLEY
DUNDEE
MATUBATUBA
LADYBRAND
PIETERMARITZBURG
THABA-
NCHU
EMPANGENI
BLOEMFON-
THABANA
VALLEY OF
GINGINDHLOVU
TEIN
NTLENYANA
1000 HILLS
STANGER
PINETOWN
DURBAN
DE AAR
GRIQUA-
LAND
E.
PORT SHEPSTONE
TRANSKEI
PORT ST. JOHNS
STUTTERHEIM
EAST LONDON
PORT ELIZABETH
RAILROADS (MAJOR)

KRUGER PARK
DRAKENSBERG

period of higher temperatures. All the veld turns green, water is abundant, rivers come to life, wild flowers bloom, the dust lies down, and the country is Africa at its best— the elevation keeps temperatures moderate, the nights remain cool, and rain, while frequent, comes in storms which quickly clear to let the sun shine again.

In contrast to the highveld, the coastal regions can be hot and oppressive. In Durban, the humidity is almost always high, temperatures are rarely low, and rain falls at any time during the year. Unlike Johannesburg, Durban has no really cool winter periods, and when the highveld experiences its dry, dusty winter, holiday-makers flood to the beaches of Durban and Natal. Many who find Durban too hot and humid consider Cape Town, with its mediterranean clime, to have the finest weather. In Cape Town winter can be moody. Rain falls, and gray days occur. The total rainfall is not heavy, however, and the amount of sunshine received is very great. Summers are dry and may be hot, but they are never as oppressively hot as those of humid Durban. Johannesburg, Cape Town, and Durban typify the three climatic regions of South Africa: the plateau, the mediterranean, and the humid subtropical area.

| *Johannesburg and the Highveld*

THE GOLDEN CITY

In all of Southern Africa, Johannesburg—which calls itself the "Golden City"—is by far the largest urban agglomeration. Yet where Johannesburg now stands and sprawls, where nearly 1,100,000 people live, there was nothing but veld less then 80 years ago. Johannesburg is the focal point of the railroad and highway network of South Africa; it is the financial capital of the country; and it is the core of the Witwatersrand and the entire southern Transvaal, a region which contains over 3 million people, or about 20 per cent of the South African population. A bustling city, growing still, Johannesburg and its huge suburbs are occupied by representatives of all the racial and cultural groups that are found in the country—African, white, Indian, Colored, Chinese, to name only some.

Johannesburg, which dominates the economic life of South Africa, owes its existence to the gold of the Witwatersrand, and the gold-mining has stimulated the development of many industries in and around Johannesburg proper. Engineering, metal-making, food-processing, and building are to be found on the Witwatersrand. Some industries have grown in response to the needs of the mines, others as a result of the consumer market of the southern Transvaal, which is the largest one in South Africa. Mines and industry drew large numbers of Africans to the "Rand," and Johannesburg became and remained the country's most cosmopolitan city. Bible societies state that they sell Bibles in Johannesburg in some 100 languages, of which many are African tongues. Xosa, Zulu, Basuto, Swazi, Nyasa, and a host of other African tribes are represented in Johannesburg's African suburbs. But it is not only the African sector of the population that is varied. Jews, Portuguese, German, French, Irish, Greek, Dutch and Belgian people, among many others, mingle with South Africans. Johannesburg cannot be called an "Afrikaner City," as Pretoria is often called, or an "Englishman's town," like Pietermaritzburg, the capital of Natal. English-speaking South Africans and Afrikaners are both represented in large numbers, and while in many ways Johannesburg is South Africa at its worst, the city is too large to be dominated by either of the two factions of the white population, and from that point of view it might be seen as white South Africa at its best. There are perhaps 450,000 white people in Johannesburg, and of the remaining 650,-000 the bulk are African.

One of the striking characteristics of Johannesburg is the wealth of the white sector of the population, as displayed in the large, beautiful homes and spacious gardens of Houghton, Parkview, Bryanston, Northcliff, and other suburbs in the north, with their golf courses and country clubs. North from the central city, Johannesburg looks as affluent as the best American samples of suburbia. Almost without exception, each home has its garage (sometimes double garage, since many Johannesburg families are two-car families) and servants' quarters behind the garage, hidden in the least obvious part of the garden. Until the recent labor shortage resulting from government efforts to relocate Africans who did not have proper passes with which to enter the city, it was difficult to find a family without at least one African servant, and the exceptions today are still very rare. Some families have two or three servants, and a few have even more.

The quarters for these servants, who are not permitted to live in the master's home, may be quite spacious, but sometimes they are mere cubicles, floored with cement and roofed with corrugated iron sheets. Yet the contrast between the African servant living in his own—if poor—accommodations in the white suburbs and the African living in the African suburbs (Johannesburg, like all South Africa, is segregated) is sharp. Although there are many white middle-class suburbs such as Parkhurst and Bezuidenhout Valley where the lot on which the house stands is barely large enough to accommodate any kind of servant's quarters, there is nowhere in white Johannesburg the crowding, squalor, and poverty of Moroka, Alexandra, or Orlando. South and east of the central city lie the great African ghettos—as they have been called—where thousands of people live in shantytown slums and makeshift dwellings, attracted by the city and the hope of finding work and making money there.

It would be unfair not to mention the efforts that are being made to eliminate these conditions and to improve the lot of the legitimate African resident of the city. Johannesburg and the Witwatersrand, however, have attracted Africans from all over the subcontinent by the hundreds of thousands, and the rate of arrival has kept steadily ahead of the capacity of the city to absorb them. In their dwellings constructed from cardboard boxes, discarded pieces of iron, gasoline containers that have been cut apart and flattened out, and other scraps, those who live in the African ghetto suffer severely during the winter, for Johannesburg's nights can be bitter when the south wind howls over the plateau, bringing with it the cold air from the higher latitudes. Many organizations attempt to help the people, and bedding, clothing, and the like are gathered each winter for distribution to those who are most in need. But whatever the sufferings of the people already there, exposed to disease and terrorized by gangs, more Africans arrive each day, hoping to obtain work, a pass—and wealth.

Even where tens of thousands of neat homes are being built to accommodate residents of the slums, resettlement does not always go smoothly, for there are people who prefer to stay where they are, because they are closer to work or because they have got used to their environment. When Africans were moved to the shiny new—but far away—suburb of Meadowlands from Sophiatown, one of the city's most terrible slums relatively near the city center, those who had been exploiting the slum-dwellers, packing people literally by the dozen into structures made many years ago to house three or four, of course objected. It was unfortunate that some sought to make political hay of this removal, which had its less pleasant aspects but which was absolutely essential for the well-being of the people most directly concerned—the occupants of the dilapidated structures. The trouble in Johannesburg and in South Africa generally is that the law which forces changes of this kind is a law repugnant to most of the people in the country. The move to Meadowlands was necessary, but it was initiated at least in part because the Africans in the slum in Johannesburg were creating a "black spot"—an island of African residence in what was to be white Johannesburg. The people who were moved from Sophiatown were moved not because they were the worse off among Johannesburg's slum-dwelling Africans but because here was an undesirable situation in the eyes of the policymakers. This sort of thing costs South Africa untold quantities of good will. Many of the Africans were genuinely pleased with their new homes but spoke bitterly about the real reason for their removal from Sophiatown.

Rich or poor, Johannesburg's homes have one thing in common: they are all "burglar-proofed." Each window that opens is guarded by a wire mesh, more or less elaborate, to prevent burglars from entering by them. This "burglar-stopper" may be a simple wire netting attached to the inside of the window frame, or it may be a series of ornate wrought-iron structures welded to the steel window frame and virtually indestructible. For would-be thieves, this is a real challenge, and one of their favorite tools is the fishing rod: an open window allows a fishing rod and line to be inserted through the mesh, and a sizable hook will bring such profits as handbags, suits with well-filled pockets, and the like. To discourage the owners of these articles from grabbing the fishing rod, it is adorned with razor blades and nails. It is, therefore, best to hang clothes in locked closets at night and to have as little loose property as possible in any room where a window is open. Some Johannesburgers report that their bedding has been stolen from their beds while they slept.

Theft is rampant in Johannesburg, and not only petty thievery. In the huge African slums, the formation of rampaging Tsotsi gangs was a natural consequence of prevalent conditions, and these gangs of teen-age Africans have proved a major headache to the overworked police. Nor is robbery solely interracial; those who believe that the African will attack and rob only a white man are sadly mistaken. African wage-earners are frequently robbed of their week's pay by African outlaws, and cases of violence involving stabbing and shooting between Africans far exceed in number those affecting whites and Africans. Car thefts, shopbreaking, and well-planned grand larceny are common in the long Johannesburg nights.

Johannesburg is a tense city. Nowhere else in Africa have so many people of different racial and cultural groups been thrown together in such a small space in such a short time. Attempts to deal with the problems resulting from rapid importation, exploitation, and urbanization of African labor often cause even worse problems.

The much-discussed pass law may be seen in this light. Under this law, no African can enter Johannesburg (or other urban places in South Africa) without a special pass which is his identification and upon which his place of work and residence are indicated. This law, established to stem the tide of influx of workless Africans into the cities, has become one of the country's most criticized regulations. There can be no doubt that it constitutes an infringement on the individual rights of African citizens of South Africa. On the other hand, the law did ameliorate the living conditions of the Africans already in the cities, although many who were there without the necessary papers were compelled to return to their territory. Those who favor the pass law question whether a man must be permitted to be miserable, homeless, and workless in a city or whether he should not be restrained from making himself a liability there, not only to the family he may have brought along, but also to the existing community, which must now try to assist him when barely able to help itself. The pass laws, however, like so many other laws which are advertised as for the benefit of the African himself, are a means of control, and this is one reason why the government will not retreat on the issue and why the police will shoot in maintaining it. The Sharpeville shooting, in which over 60 Africans were killed in 1960, was the result of a mass objection on the part of Africans to the law and nervousness on the part of a small police party finding itself faced with thousands of hostile people. Africans will often, when asked why they object to the pass law, point out that there is no similar law to stem the tide of white entry into the cities. There is, of course, more control now also over the white population than there used to be. Everyone now must carry an identification card, but there is no equivalent for whites to the pass law for Africans. It would be false to suggest that there are no undesirable white elements in the cities of South Africa, and the African claim that the pass law is discriminatory is easily justified.

Johannesburg is a bustling as well as a troubled city. Nowhere else in Southern Africa does life move with such speed. Yet Johannesburg does not have the environment to stimulate sustained hard work. Between 5,000 and 6,000 feet, the air is rarefied, and visiting soccer and rugby teams will attest to the effects of this condition upon sustained effort. People feel a need to go to the coast once a year, and Johannesburgers blame many of their problems on the altitude. A day of walking and shopping in the center of the city is more tiring than in any other in the country. Everyone looks tired, and rural folk talk of the irritability of the *stadsmense* (city people). Low production figures, quick exhaustion, mental slowness—all these are blamed on the "mile," the mile separating the city from sea level. One might in addition mention the dryness, dustiness, and monotony of the winter, which is often accompanied by water shortages. For mental work, Johannesburg in winter is most unsuitable. Johannesburg happens to be one of the two largest cities in the world to be located on a divide, and water rationing is often necessary, partly because of the amount of spraying the owners of large gardens do when the drought prevails.

Johannesburg, a city thrown together in a few years, is planned along modern lines. A map of the central business district shows neat square blocks separated by some fairly wide streets, such as Commissioner Street, and some very narrow ones, like Joubert Street. The central city is fairly tightly wedged in by the mine dumps in the south and the railroad station and lines in the north, and it has expanded vertically, creating a skyline which is formidable in South African eyes, though not, as some would have it, comparable to, for instance, Pittsburgh or Kansas City in the United States. There are buildings over 20 stories high, and the downtown streets are stone canyons similar to those of many American city centers, but Johannesburg has retained a small-town character which it cannot seem to throw off. The narrow sidewalks are crowded with shoppers, but store windows are unimaginatively decorated. In Eloff Street, where until recently the old trams still ran, adding to the everyday traffic confusion, a fascinating cross-section of the South African people can be observed. Here, office workers rub shoulders with Afrikaner farmers in town for the day, and African delivery boys and shoppers from the African suburbs mix with new arrivals from Swaziland, Basutoland, or Zululand, some still in their traditional dress. Indians and colored people mingle with white clerks and uniformed girls from the large department stores. Everyone is in a hurry. The streets are noisy, traffic jams are common, and parking places at a premium. In some ways a brief look at Eloff Street and a glance at the main street at Atlanta, Georgia, might leave similar impressions. With one car to every four white people, the central business district, poor in vertical and underground garages, is ringed by huge parking lots which on any weekday resemble those outside Chicago's Soldier Field at football time.

With its legacy of modern planning, Johannesburg has not managed to attain any real character architecturally. The downtown area is an agglomeration of buildings of all kinds, from the unbelievably ugly City Hall to quite imaginative structures. Almost always, there seems to be a building boom. Buildings are replaced and constructed at a rate which is said to be among the highest in the world. Africans in the building trade are paid better than many in other jobs, and this may be one reason that their rate of work is high. Much is said of the laziness and slowness of Africans and how their usefulness in the gold mines is limited because of their low rate of output, but the African construction worker in Johannesburg could hold his own in competition with American and European workers. As a result of the high rate of replacement, the face of downtown Johannesburg is improving, and were it not for the very dirty air which stains buildings faster than anywhere else in Southern Africa, the city might in time come to look quite attractive.

The morning and evening traffic rush into and out of the terribly congested central city resembles the scramble in many American cities. Johannesburg has no subway, and the city is spread so far and wide that the proposition to construct one has not been found tenable. From the north come thousands of cars, and from the south, east, and west many Africans flood into the central city and its immediate environs, by train, bus, and bicycle. Incredibly crowded buses (for Africans only) speed down Louis Botha Avenue, while the characteristic English double-deckers, motor and trolley, bring in commuters from the white suburbs. For the working day, black and white rub shoulders, and white money and power and African muscle and sweat combine in the common economic effort. Then everyone returns home, each to his section of the town. In the morning the Afrikaner reads his *Transvaler,* and his English-speaking colleague buys the *Rand Daily Mail.* In the evening, both buy the *Star,* although there is an Afrikaans evening paper called the *Vaderland.* There is also a widely read African press.

Johannesburg has a good library (including an excellent Africana collection) and some fine museums, but the city is something of a cultural desert. It has never sustained a really successful repertory company. Plays are offered, but they do not run very long. In the late forties and early fifties, there was the Johannesburg City Orchestra, a municipal organization with a small but loyal following, which blossomed for some years until its costs began to bother the Johannesburg City Council. Various efforts were made to make the orchestra pay for itself, promenade concerts were offered, and several well-known guest conductors were invited. Eventually, amid considerable upheaval in the city's cultural life, the orchestra was disbanded, and a new orchestra formed of members of the orchestra of the South African Broadcasting Corporation and the former Johannesburg City Orchestra. The merged group, known as the

South African Broadcasting Corporation Symphony Orchestra, does not compare favorably with most of the semi-amateur orchestras of American cities smaller than Johannesburg. When several musicians who had left good positions in Europe to play in the ill-fated Johannesburg City Orchestra were not hired for the combined orchestra, the South African Musicians' Union got an international reputation roughly comparable to that of the South African Government in international political circles. Overseas chamber groups, famous artists, and opera companies take care not to stay long in Johannesburg, and one of the embarrassing aspects of musical life in the city is the quality of press commentary. There is a page in the *Star* under the editorship of a music critic whose writings have to be seen to be believed. Not so long ago this newspaper was sued successfully by a well-known chamber ensemble which was criticized for the performance of a work which was never played!

After dark, Johannesburg dies. Except under the bright lights in front of a few movie houses, streets are nearly deserted when darkness comes, and in South Africa this is never very late. Much of Johannesburg is dangerous after dark, and if it is not necessary to go in the street at night, people usually don't. The theaters cause a brief flurry of activity around 8:00 P.M. and 11:00 P.M., if any plays are running, but they of course are for whites only. During the night, Johannesburg's Flying Squad police keep vigil in the deserted city, and the morning newspaper will show the night's crop of lawlessness. It is difficult to envision anything quite so dismal as the central city of Johannesburg after the evening commuters' rush. Few store windows are lighted, and only the hum of a trolley-bus lends variety to the scene. There are some nightclubs (the number has increased recently) and a few restaurants open at night, but South Africans who compare Johannesburg to New York, as many are likely to do, do not know what they are talking about.

Sunday, also, makes of Johannesburg's downtown area

an empty shell. Sports events are held on Saturday afternoon, and the big night for party-goers and theaters, as in the U. S. A., is Saturday evening. On Sunday, all theaters are closed, and stores are closed with exception of some restaurants which may sell essentials such as milk, bread, and the like but are prevented by law from selling anything else. Sports events before a paying public may not be held, although professional soccer has made some inroads into this rule in recent years. A look at the roads leading from Johannesburg to the Hartebeestpoort Dam, Magaliesberg, or Vereeniging will convince one that all Johannesburg is on its way out of town; South Africans will drive any number of miles to enjoy a picnic. There is nothing to do in town except sit at Zoo Lake.

The people who travel into the area surrounding Johannesburg get a look at the heart of South Africa. From tens of miles away, the huge mine dumps south, east, and west of Johannesburg remain in sight. These yellowish white piles of sand, which are made up of crushed ore that has been worked for gold, are a headache to the city and this part of the southern Transvaal. Any strong wind blows great amounts of sand into the city from them, and they have been known to bury unfortunate climbers. Efforts have been made to plant grass on these dunes, which are hundreds of feet high, and a plan was even forwarded to construct a large hotel on top of one of them. Between the piles of sand are the characteristic shaft-heads of the mines, with the huge wheels of the elevators, and a more or less haphazard arrangement of structures. Johannesburg's mines show signs of failure to yield profitable quantities of the mineral. In addition, there are competitors on the equally rich fields in the Orange Free State, and it is not surprising that the government is encouraging industrial diversification. Water supply, in this connection, remains a problem.

Southward, the Transvaal extends to the Vaal River on which Vereeniging lies, and northward from Johannesburg lies Pretoria. From Vereeniging through Johannesburg to

Pretoria is the north-south axis of the southern Transvaal. Eastward and westward from Johannesburg lie old mining areas of the Rand. The east-west axis of the Rand is also the backbone of the southern Transvaal. With the decrease in importance of mining and increasing industrialism, the Rand towns are taking on a new character. Those to the west of Johannesburg are becoming suburbs of the Golden City. Johannesburg itself is partaking of the change of affairs on the Rand in general, and 90 per cent of the white workers in the city are now in industry and manufacturing instead of mining. It is, therefore, incorrect to see the Witwatersrand, Johannesburg, and its surroundings, as a mining region with a limited future dependent upon the remaining workable ores. Johannesburg has rapidly changed from a mining town to a city with a great deal of diversification, and industry takes over where mines are abandoned.

STEEL AND NATIONALISM

While industry is spreading along the east-west axis of the Witwatersrand, with the east Rand forging ahead because of its more favorable location with respect to topography (the area is less hilly) and communication lines (it is closer to the ports), the north-south axis does not display such remarkable progress. Vereeniging, Johannesburg, and Pretoria are linked by railroad, and along this axis, communications and administration are the main activities, so that there has not been the mushrooming growth of the Rand. Nevertheless, Pretoria ranks today not only as the capital of South Africa but also as the home of the oldest steel mill in the country, ISCOR (Iron and Steel Corporation). Whereas Johannesburg's incredible growth was based on gold, Pretoria is one of these few cities in the world to find itself located close to coal deposits, iron ore, and limestone, without a shortage of water. The coal comes from the east, where along the great road to Lourenço Marques lies Witbank, and the iron ore comes from the northwest, at Thabazimbi. As the first of a series of steel mills, ISCOR has proved of tremendous importance to South Africa in both a material and a psychological sense. In all subsaharan Africa, only South Africa and to a lesser degree the Central African Federation have sizable steel mills. South Africa's first steel works reached the stage of production in 1934, after a great struggle to assemble sufficient capital. The plant was then capable of producing rails, structural sections, and some other products for use in industry. This encouraged the expansion of the country's railroad network. Particularly in 1939, with the onset of the war, the South African steel industry proved its worth to the country in this and other respects. When demands upon the local industry increased greatly, a site near Vereeniging was chosen for another steel mill, because of the available power supply from a new power station, the local water supply from the Vaal River, and the growth of industrialization in the city in general. Today, Vanderbijl Park, the steel suburb at the southern end of the north-south axis of the southern Transvaal, is a showpiece of planning and a vital part of South African industrial development. The industry supplies well over 70 per cent of the requirements of the country, has stimulated engineering and allied industries, and affords work for some 100,000 persons.

Whereas the east-west backbone of the Witwatersrand is one continuous string of towns, a virtually uninterrupted built-up zone, along which mines, industrial plants, and residential areas vie for space, the north-south belt connecting Pretoria and Vereeniging is less developed. Travelling from Springs in the east to Krugersdorp in the west along the Rand is like going along a practically continuous city street, but going the longer distance from Vereeniging to Pretoria one crosses large areas which are still open veld. Pretoria and Johannesburg are growing toward each other, however, and in the foreseeable future the outskirts of one may be within less than ten miles from the other.

Pretoria, the capital of the South African Republic from 1860 to 1900, has grown rapidly without the stimulus of gold finds. Had its growth not been overshadowed by that of Johannesburg, it would be as famed for expansion as is the Golden City today. It is no less remarkable, for Pretoria has had slight significance other than as an administrative center. Other places in South Africa with similar qualifications have failed to show Pretoria's vitality. For rapid urban development, in Africa one naturally looks first at the mineral finds that may be the cause, or the location of the port with reference to its hinterland in the case of a coastal town.

The city lies on the southern flanks of the Magaliesberg Range, and, as Cape Town is the gateway to Southern Africa, so Pretoria is the gateway to the northern Transvaal where in terms of size it has no competitors. North of Pretoria, only Pietersburg is a town of any appreciable size, and it is about 200 miles away. Pretoria is the focal point for a huge area that includes some rich agricultural land, some important mines, and a sizable population. Although part and parcel of the southern Transvaal, Pretoria is stimulated in its development by the products of regions far beyond, and railroad links with the north, east, and west help to funnel these materials to the capital. As the seat of the government (Pretoria is the administrative capital of South Africa, Cape Town is the legislative capital, and there is an annual migration on the part of government officials) the city possesses some obvious advantages.

Compared to Johannesburg, Pretoria is a beautiful city. There are several architectural landmarks. Against the hills stands the vast Union Building, overlooking some of South Africa's most beautiful gardens and lawns. Another famous structure to be found just south of the city is the Voortrekker Monument, opened in 1949. This huge building, located on a hill, is visible from as far away as Johannesburg, and while its aesthetic qualities are a matter for debate, no one denies that it is striking.

The Voortrekker Monument is of great significance in the South African's (especially the Afrikaner's) life. It is the ultimate result of the vow to God, made by Cilliers before the Battle of Blood River in 1838, that should the trekkers be granted victory, a temple would be built in gratitude. Some might express surprise that 111 years should elapse before something was done about this promise, but the result is at least impressive. The monument has been built of gray granite in a style more or less adapted from that of Zimbabwe and the Zulu. The monument is encircled by a wall depicting 56 covered wagons such as those used by the trekkers. At three corners of the monument rise huge granite statues of the well-known trekker leaders Retief, Pretorius, and Potgieter. In the fourth corner is a characterization of the Unknown Pioneer.

Within the temple the "flame of civilization" burns beneath a granite table which bears the inscription "Ons vir jou Suid-Afrika" (We for you South Africa). The "flame of civilization," which flickers in the cellar beneath the chamber, is the symbol of the white civilization which was brought to the dark interior by the trekkers. The granite table and its inscription are located in such a manner beneath a small opening in the dome which forms the roof that on each December 16 a ray of the noon sun falls on it. The flame, inscription, encircling wagons, statues, and elaborate sculptured impressions of the Battle of Blood River are all holy to the Afrikaner people. Each December 16, huge throngs assemble at the monument and gather in the adjacent amphitheater for a day of prayer and thanksgiving. Until the erection of the monument, December 16 was a holiday known as Dingaan's Day; today it is known as the Day of the Covenant.

When, in 1949, the Voortrekker Monument was opened, some 250,000 people came to Pretoria and camped at the site. The occasion was fascinating to witness and full of intentional and unintentional symbolism. The morning of December 16, a service was held in the great amphitheater, and among the distinguished speakers were D. F. Malan, then the Prime Minister of South Africa, and J. C. Smuts, his defeated opponent, then leader of the Opposition.

Malan spoke of the gains of the country and of his people, and the old statesman was at his best as he brought the throng to tears with an account of the trials and tribulations of the trekkers, their unswerving faith, and the importance of their success. Even when someone made a rush at the stage from which the Prime Minister was speaking, he hardly faltered. Following Malan, Smuts spoke at length to an audience which by and large saw him as a traitor, a man who had fought on the Boer side in the Boer War, who had battled subsequently for the unification of the people in his country, and who had then, in the view of many, sold out to the English. It was an unforgettable experience: Malan, self-assured, in good health, well spoken, cheered as a victor by the victorious; Smuts, speaking slowly and haltingly, a man facing defeat and a lost cause. It was a warm day, December being a summer month in South Africa, and Smuts spoke late in the proceedings. His address was an anticlimax to the many who had come to savor the excitement of a resurgent nationalism, and hundreds filed noisily out of the amphitheater. Smuts was once interrupted by the master of ceremonies, who announced that "never will you have a chance to hear this again! Stay to listen!" But the exodus continued while Smuts spoke, and the tired crowd retreating to the comfort of a meal and shade represented a people turning its back on reason and moderation.

What happened at the Voortrekker Monument that December 16 was a sign of things to come in South Africa. Superficially, the scene was tranquil enough. Speakers had addressed the people, many of whom wore traditional costume, in both official languages. The words "unity is strength" were uttered frequently. It is, by the way, interesting to note how this "unity is strength" was translated into Afrikaans, where it is "eendrag maak mag." The latter, literally translated back to English, reads "unity makes might." In Afrikaans, there is a significant difference between "mag," meaning force or might, and the word that should have been used, "krag," which literally means strength or power. It should have been "eendrag is krag," but "mag" is nearer reality. But while unity among the whites was stressed, this was the Afrikaner's day, and the Afrikaner's only. All too often the memories of what happened to Afrikanerdom after Blood River at the hands of not the Africans but the English were conjured up, and the Day of the Covenant is not treated by all English-speaking South Africans with the same respect as it is by Afrikaners. The lager spirit, the heroics performed by the pioneers within their circle of wagons, is not dead among the Afrikaners of today. That circle of 56 wagons around the monument, the granite images of trekkers, the flame of civilization, the ray of sunlight—all these reflect much of the Afrikaner philosophy of today. In a way, the isolationism, chauvinism, and conservatism, illustrated by the gray squareness of the temple, have intensified among the Afrikaner people, rather than waned. South African nationalists tend to see South Africa more than ever as an embattled bastion reminiscent of the trekker lager in a hostile world and a savage continent.

Beside the steel works and the Voortrekker Monument, Pretoria has many other noteworthy aspects. The capital of the old South African Republic, the residence of President Kruger, now a museum, can still be found on Kerk Street, and some old buildings survive from those days. Church Square, the center of the city, is ringed by impressive buildings and makes an infinitely more pleasant appearance than does the congested area in front of and behind the Johannesburg City Hall. Johannesburg really does not have a central square, and what it misses is well shown by Pretoria. The Department of Justice is housed today in the Palace of Justice, a large and ornate structure dating from Boer War days. In the center of the square stands an imposing statue of President Kruger, moved not long ago from in front of the railroad station. Church Square is one of South Africa's major crossroads. The Great North Road, the road to the eastern Transvaal, and the road to the west all lead to the square, which is busy day and night.

In common with several other South African cities, Pretoria has a modern, rectangular city plan. For this, the city has Du Toit to thank, for his plan of the city, dated March 2, 1859, has largely been followed. The Square was a part of the plan, and today Pretoria boasts the longest street in South Africa and some beautiful, wide avenues. Through the years, the Kerkplein has seen momentous events take place. This is where the troops assembled for battle in the Boer War, where festivities were held by the farmers, and where open-air church services took place to mourn dead warriors.

Pretoria is known as the Jacaranda City. In October, over 50,000 Jacaranda trees bloom in colors which range from purple to violet depending upon the time and light of day. The streets are lined with these unusual trees, and their blooming is the signal for a carnival, the choosing of Jacaranda Queens, floats elaborately decked out with flowers of all kinds, and a generally festive atmosphere. There are other cities in South Africa which have planted these trees and where they have grown successfully, but nowhere is there so much beauty as in October in the capital. In addition to the Jacarandas, Pretoria has some of the country's most magnificent public and private gardens, and in places the town resembles a sea of flowers, particularly during the beautiful spring and fall. The city has the advantage of lying in a slightly hilly area, so that beautiful views over the area can be had from many homes, and from the city, the slope against which the Union Building lies is visible almost everywhere.

Lying somewhat lower than Johannesburg, Pretoria is warmer. The city is less rushed, and life goes at a pace more familiar in Africa than the one displayed by the Golden City. The city center, the area around the Church Square, is impressive but not as much so as Johannesburg's central business district. There are fewer high buildings, though there are some very modern ones, and the streets are less like gray stone canyons than those of the nearby rival.

The white population of Pretoria has risen from 70,000 in 1938 to about 160,000 in 1960, an increase of 128 per cent. Johannesburg, by comparison, over the same period saw an increase in white population of about 40 per cent, Cape Town about 28 per cent. The other city with rapid growth is Durban, which over this period had an increase of 75 per cent in its white population. The total population of Pretoria is now probably about 375,000.

Unlike Johannesburg, Pretoria is an Afrikaner-dominated city. Afrikaans and Hollands—there are a great number of Dutch immigrants here—are heard almost exclusively. Since it was never Pretoria which drew profiteers to the wealth of the southern Transvaal, this is not surprising. From Pretoria northward stretches Afrikaner South Africa at its most intense, perhaps even more so than the Orange Free State, which has been affected to some degree by proximity to the more moderate Cape and recently by its own invasion of foreigners as a result of the discovery of gold deposits. In the Transvaal, to the north of the capital, survive the spirit and philosophy of the republics, and Pretoria reflects them. Possibly because this spirit is what it is, the city is not a cultural haven even of Afrikanerdom. It has an insignificant press, and the foremost Afrikaner Nationalist daily paper is still printed in the Cape. Pretoria is poor in live theatrical performances, does not sustain its own permanent professional company, and is blessed only a few times each year with a visit by the S.A.B.C. Symphony Orchestra, failing itself to produce even a semi-amateur orchestra. Surprisingly few of the great names in Afrikaner literature, painting, music, and other art forms hail from Pretoria, although the historical and political heritage of the people is evident everywhere and make for an atmosphere laden with intense national pride.

Pretoria shares with Johannesburg the evils of rapid urbanization, and the impression many visitors from the

Golden City have that this is a "safer" city, at night for instance, are mistaken. Near Pretoria is a park and picnic area called the Fountains. This has long been a favorite place for Pretoria couples out late at night, and many have come to regret their choice. For a long time, one or more Africans terrorized this place, and attacks on people in parked cars were so frequent that the police advised against visiting the area after dark. Several fatal and near-fatal slashing attacks were made by an African who came to be widely known in the sensation-hungry press as the Panga-man. Like Johannesburg's homes, houses in Pretoria have to be burglar-proofed, and there is little reason for feeling secure when out on foot after dark. From the point of view of the Afrikaner residents, Pretoria does have the advantage of a more favorable white-black population ratio. The African suburbs are smaller, form less of an attraction than do those of Johannesburg, and the shantytown problem is less severe.

OUTSIDE THE CITIES

Johannesburg and Pretoria rise head and shoulders above the remainder of urban development in the Transvaal. Westward lies Rustenburg, more or less at the western end of the Magaliesberg Range. North, near the geographical center of the Transvaal, is Pietersburg, and the eastern Transvaal is dominated by Nelspruit. Thus the center of gravity of the Transvaal is very much in the south. Peculiarly, there is something unique about each of the fringes of the province. In the north, the Zoutpansberg runs east-west for great distances. This is not a very prominent range, although the approach from the flat highveld in the south is impressive. The range separates the area between it and the real boundary, the Limpopo River, from the core of the Transvaal, the area centering on Pietersburg. After the long and beautiful drive north from Pretoria, crossing via Wiley's Poort through the Zoutpansberg range is like leaving the country, and there is something detached about the plain to be crossed before Messina and Beit Bridge are reached. If one travels westward beyond Rustenburg, the increasing dryness of the country in the approach to Bechuanaland, the failure of the lush highveld grasses, and the sparseness of population again create the impression of separation from the hub of the province. Eastward, there is something of a parallel again; the larger part of the border with Moçambique runs through the Kruger National Park, which occupies the rich eastern Transvaal from the Limpopo River in the north to the Crocodile River in the south and covers over 8,000 square miles, an area equivalent to the state of Massachusetts. South of the Kruger National Park, which is a fitting monument to the man who took the first steps in setting aside land for wildlife preservation in the country, is Swaziland, so that the eastern Transvaal possesses a narrow occupied corridor along which the railroad and road to Lourenço Marques from Johannesburg run. Again, it is necessary to cross a mountain range to enter this region, and it is separated from the area dominated by Nelspruit and thus from the rest of the Transvaal. In addition, the traveler going eastward to the coast descends the escarpment, and the country changes from the grassy plains of the highveld to the beautiful middle and lowveld with its parkland savanna.

The western, eastern, and northern Transvaal each make their contribution to the national economy, and the contribution is largely agricultural. There are some mines, producing chromite, platinum, asbestos, copper, and other minerals, but the aspect of each area is agricultural, and the mines do not dominate these regions as do those of the Witwatersrand in the southern Transvaal. The western Transvaal in the vicinity of Rustenburg is a pleasant, thriving citrus-fruit and vegetable-growing country, but farther west the country is good only for corn and cattle. West of Pretoria lies one of the favorite picnic spots for both Johannesburgers and Pretorians: the Hartebeestpoort Dam. This and the Loskop Dam east of Pretoria are two of the several dams built to tide the Transvaal over the long dry winter. The lake created by the Hartebeestpoort Dam, pic-

turesquely located between the cuestas of the Magaliesberg
bankeveld, is a large body of water but has nearly run dry
on some occasions. Unfortunately, as elsewhere in the region, the dreaded snail-carried disease of Bilharzia lurks in
these waters, so that people must be warned not to swim.

The Great North Road from Pretoria to Beit Bridge on
the Limpopo River and on into Southern Rhodesia is one
of the country's most interesting highways. It leads through
the northern Transvaal, providing a sample of the scenic
variety of this region as the Magaliesberg range, Waterberg plateau, Pietersburg plain and Zoutpansberg range are
crossed. Immediately after leaving Pretoria the road cuts
through the Magaliesberg, and then begins the 60-mile
stretch to Warmbaths. This part of the road goes over flat
country, typical bushveld—the area is called the Bushveld
Basin. It is one of South Africa's geological wonders, the
largest lopolith intrusion of igneous rocks in the world, so
vast as to tilt the rocks to the south into the range now called
the Magaliesberg. From east to west, this intrusion is measured in the hundreds of miles, from north to south in dozens.
These rocks are the source of the chromite and platinum
South Africa produces, and along the rim of the basin, mines
are located. Warmbaths lies on the southern slope of the remains of the overlying strata, the Waterberg. The plateau
of the Waterberg rocks is characteristically red colored,
rather dissected, and the short trip through the valleys is a
pleasant interlude after the long haul over the bushveld.
Nylstroom lies on the northern flank of this section of the
plateau, and it is the birthplace of the late Prime Minister of
South Africa, J. G. Strijdom. Through this town runs an insignificant little stream, the Nyl (Nile), so called because
when the trekkers first arrived here the stream was a miles-
wide surging body of water, and the trekkers thought they
had reached the Nile River! To South Africans, who know
the variability of their rivers, this is nothing unusual, but the
first impression of the little gulley today makes the mistake
almost unbelievable.

North along the Great North Road lie some typical Afrikaans *platteland* (rural) towns, such as Naboomspruit and the picturesque Potgietersrus, home of a high school of excellent repute. All the while, the road runs through relatively flat bush country, diversified occasionally by a remnant of plateau rock or a granite hill, sometimes called *koppie*. After Potgietersrus the country remains attractive for some miles, and then begins to turn barren. The traveler approaches Pietersburg, visible from a dozen miles away across the treeless highveld plain.

Pietersburg is not the most attractive town in the Transvaal, but it is growing rapidly and has been described by some observers as a miniature Pretoria. It is the junction of the road leading to the eastern Transvaal and is a good example of response to the requirements of a region. To the north are over seventy miles without any urban development at all, to the south Potgietersrus is hardly competition, and Pretoria is some 170 miles away. To the west there is very little development, and no main road leads directly westward out of Pietersburg. Pietersburg is on the railroad leading from the border at Beit Bridge (where it stops) to Pretoria, Johannesburg, and south. For the farmers of the immediate neighborhood, it is a market, and for a large surrounding region it is the focal point.

Between Pietersburg and Louis Trichardt the traveler finds typical highveld country, but soon afterward he crosses the Tropic of Capricorn, enters tropical South Africa, and the elevation begins to fall. Some magnificent *inselbergs* (island mountains) of granite rise above the flat surface of the plateau. From a dozen miles away, the south-facing scarp of the Zoutpansberg Range comes into view, at its foot Louis Trichardt, named after the famed trekker. By contrast to the granite plain immediately south of the range, the mountains are quite moist, and 80 inches of rain is recorded in some of the valleys. The Zoutpansberg make a beautiful series of mountains, densely bushed in some places, afforested in others, and bearing bananas and citrus fruits elsewhere. Soils, derived from basalts or sedimentaries, are red colored, and the faulting in the rocks has produced numerous springs. Yet the northern Transvaal, in a way, is the forgotten country of the province. Only recently has the road from Louis Trichardt along the southern foot of the range to the east been provided with an all-weather surface, and the installation of electricity supply lines, long promised, is slowly gathering momentum. The northern Transvaal suffers from a labor shortage, and white farmers have trouble to bring harvests in as a result of the lack of permanent help. Compared to the thriving eastern Transvaal, the relative lag in development here in the north is surprising.

The Great North Road crosses the Zoutpansberg through one of the many beautiful gaps (poorts) in South Africa's mountains. This one, Wiley's Poort, is in places just wide enough to permit the passage of one truck, but on the whole the road is in remarkably good condition. Like other poorts in South Africa, Wiley's Poort becomes hazardous when the heavy rains fall, and despite elaborate channels to lead away excess water, the roadway is sometimes blocked by landslides, so that an auxiliary road, involving arduous driving through miles of deep red mud, must be used. South Africans are patient in this respect, and know that even with such inconvenience they enjoy the best road system in subsaharan Africa.

North of the Zoutpansberg, the country continues to fall as the Limpopo River is approached. Baobab trees, their exotic shape a reminder of the wildness of the region, dot the countryside, and there is a good deal of wildlife left here. Messina, where South Africa finds some of its copper, lies in this region, and cattle-ranching of a precarious kind is attempted. The Transvaal north of the Zoutpansberg is a world apart, hot, dry, dusty, unchanged, isolated in spite of the main road, separated in more than one way from the rest of the country.

The eastern Transvaal, to shift locale, is full of variety, progress, prosperity. Of course this is partly due to location. Not only do a major road and railroad, connecting the Rand and Pretoria with Lourenço Marques, run through this area, but the region is nearer the outside world, within easier reach of coast and capital, and benefits from a considerable influx of tourists. The eastern Transvaal is varied in that instead of a slow drop in elevation there is the step-like, rapid descent from the 5,000 foot plateau highveld via the middleveld to the lowveld, causing the grasses to give way to densely forested river gorges (*kloofs*), bushy savanna, and warm, humid lowland. Eastward from Pretoria, after skirting the southern flanks of the Magaliesberg and passing through coal-mining Witbank with its model African suburbs, then through Middleburg with its large Asian population, the highveld ends at Machadodorp, from where the descent to the middleveld is begun. After the enervating flatness of the plateau, the green valleys leading toward Lydenburg or Nelspruit beckon like an oasis. Waterfalls, large and small, can be seen cascading off the faces of the scarps. Water collects on the valley floors, and cultivated fields begin to appear. Much of the eastern Transvaal's available land is used. The hillsides are covered with the country's most extensive forests, and the valleys grow citrus, avacados, pineapples, grapefruit, and a host of other fruits.

This country boomed when in the 1870's gold was discovered in the area around Barberton, called the Barberton Mountain Land. Barberton grew to become the most important town in the eastern Transvaal, and discoveries were made elsewhere, in the vicinity of Graskop, Pelgrims Rest, and throughout the region old mines remain to attest to the activity that occurred. The mines soon began to fail, and the Witwatersrand drew many of the miners, but the good qualities of the eastern Transvaal had been recognized. The whole region benefited from the rise in importance of Johannesburg and the consequent establishment of communication lines between the highveld and the coast.

The region benefited also from the foresight of President Kruger. In 1898, shortly before the outbreak of the Boer War, he proclaimed that 1800 square miles of the country between the Crocodile and the Sabi Rivers would be set apart for the preservation of wildlife. Thanks to the untiring efforts of Colonel Stevenson-Hamilton and a group of dedicated preservationists, the sanctuary survived poachers, land-grabbers, waves of disease which nearly exterminated the game, lack of funds, and apathy. Between 1903 and 1926 Kruger Park grew through various additions, land loans, purchases, and gifts, but World War I reduced the staff and brought renewed invasions of poachers. Anti-preservationists were for a time successful, and in 1920 coal-mining and cattle-ranching was actually begun within the present confines of the park. In 1924 a change in government brought preservationists to power, and the settlements within the park were bought out. In 1926 the National Parks Act removed the danger that the Park, which had been instituted by proclamation, could be terminated by proclamation also. Since then, Kruger Park has become the world's most famous wildlife reserve. Farms given or sold to the National Parks Board account for the enlargement of the Park to its present size. It is over two hundred miles from north to south and averages over 40 miles in width. Three of the major entrances to the park lie in the eastern Transvaal. The most commonly used entrance is the Numbi Gate, which is about 40 miles north of Nelspruit. Near this gate lies Pretoriuskop, a camp open all year round. It is hoped that more camps will be opened throughout the year, but currently the danger of malaria and other problems keep the larger part of the Park closed from mid-October to early May. The two other entrances are at Malelane and at Crocodile Bridge, both along the southern border of the Park, and they are reached via the Pretoria–Lourenço Marques road. Whoever enters Kruger Park in the south must travel through the eastern Transvaal, and whoever travels through the eastern Transvaal will fail to resist the roadside stalls selling oranges, handcraft articles, and countless other items.

It is impossible to exaggerate the positive aspects of Kruger Park even when one remains cognizant of the wildlife reserves elsewhere in Africa. Kruger Park is visited by tens of thousands of visitors from all parts of the world annually, and the most popular winter camps are booked solid long before the winter season starts. South Africans themselves are enthusiastic over their national parks, and this is one aspect of the country which everyone will heartily approve. It must be realized that Kruger National Park lies in an area which is by no means poor. The restraint shown by farmers adjacent to the Park in dealing with the animals is remarkable, whether enforced by law or not. Now the Park is beginning to pay its way, after being a considerable liability for decades.

The focal point of the eastern Transvaal is Nelspruit, one of the most pleasantly located places in all South Africa. The development of this town and its environs offers an excellent illustration of the impetus provided to urban as well as rural areas by the establishment of communication lines. Nelspruit's first building was built in 1884, when the railroad linking the present town with Lourenço Marques was begun. The first inhabitants of the place were people involved in the building of the railroad, which was completed in 1892. In 1900 the settlement, then consisting of a few scattered buildings, was briefly the capital of the South African Republic during the last phases of the Boer War. In fact, Kruger's final proclamations came from Nelspruit, and it was from here that he departed for exile in Switzerland. Because of political unrest, the advantage of the railroad was not yet felt by the area for which Nelspruit is today the focal point; actually, Barberton, father to the south and the center for the gold mines of the Barberton Mountain Land, had grown more rapidly.

The Boer War and the competition of Barberton were not the only factors which inhibited progress in Nelspruit's area. Although the area has a pleasant climate with a rainfall of between 30 to 40 inches annually and none of the cold spells of the highveld, it paid for its climatic advantages in terms of insects. Malaria was rampant here, and it was decades before this danger was finally controlled. Another problem was water supply. The underlying granites and the variability of the flow in the Nels and Crocodile Rivers do not make for reliable sources, and money for dam development or drilling was scarce. There was a tank, established by the South African Railroads, from which the citizens of the place often took water illegally. Even though the town was planned in 1905 and development in the area took place at a considerable pace, it was not until 1926 that a more or less modern water distribution system was put into operation. Throughout this period, Barberton continued its domination of the lowveld region in the eastern Transvaal, but the increasing importance of the highveld, Witwatersrand, and Johannesburg and the improvements of the harbor at Lourenço Marques made the railroad ever more important. The advantages of proximity to this artery of transportation were reflected by the pattern of settlement, and Nelspruit began to take over the position of focal point of the region. In 1921 there were less than 125 white people in the town, and by 1931 only about 350, but in the same immediate environs, an agricultural research and experiment station had been established, a hydro-electric plant was nearing completion, and some thriving farms had been established.

Nelspruit is in 1962 within 13 hours by good trains from the markets of the highveld; roads of high quality and railroads radiate to Barberton in the south, White River in the north, and points elsewhere in the tributary region; and with a total population of just under 20,000 (over 4,000 whites) the town has become the thriving center of one of South Africa's most prosperous agricultural regions. The main road to Lourenço Marques passes through mile upon mile of citrus groves, and a host of other subtropical fruits are grown. Rice, coffee, and tobacco are among the other products, while this is also good cattle country. In the immediate vicinity of Nelspruit grow South Africa's largest stands of timber, and among the industrial activities are timber-processing and the manufacturing of the crates in which agricultural products are shipped to the market.

Food-processing plants have also been established, and canned products are exported. In addition, Nelspruit has become the distribution center of fuels for the entire eastern Transvaal, again because of its location on the main railroad.

Whether the traveler approaches Nelspruit from the north along the road from White River or from the east on the way from Lourenço Marques, the site of the place, the beauty of the surrounding hills, the proximity of Kruger Park, the green beauty of the farms—all these make for a unique atmosphere. Few places in South Africa display such prosperity as this. Unlike many of its platteland counterparts, the town of Nelspruit does not know dusty dirt roads, water shortages, or electricity problems. The town's outlay is imaginative, there are good schools, and one sees few of the corrugated-iron shacks that seem to spoil so many South African towns. Progressive yet picturesque, Nelspruit and its environs pleasantly contrast with the endless monotony of the windswept highveld.

The Transvaal as a whole is South Africa's most populous province. Although there are more Asiatics in Natal than in the Transvaal, and more Colored people in the Cape Province, the two dominant population groups are most numerous in the Transvaal. The permanent African population is some 4,200,000, while there are about 1,470,000 white residents in the province. In addition to the permanent African population there are always thousands of migratory laborers in the Transvaal, largely in Johannesburg and on the Witwatersrand, who come not only from other parts of South Africa but also from as far afield as Moçambique, Basutoland, and even Nyasaland. Approximately 1,350,000 Africans live along the Witwatersrand. The remaining 3 million occupy the African reserve areas of the province and are also found on the farms owned by whites and in the outlying sections of country towns.

Notwithstanding the degree of urbanization of the African in South Africa, the character of the tribal areas has survived, and in the Transvaal as elsewhere there are regions associated with certain individual African peoples. There are innumerable tribal names, some of which are not spelled consistently, and extensive study is required to gain a full appreciation of the situation. A few of these tribal groups have attained greater fame than others. In the Transvaal, the core of the province is occupied by a body of peoples called the Northern Sotho. They possess much of the land around and particularly east of Pietersburg. They show a preference, in building their dwellings, for the granite hills which rise above the highveld plateau, and clusters of them can be observed among the great boulders. Like many other Bantu peoples on the plateau, they keep cattle and goats and grow corn. Located also in the central part of the Transvaal are the picturesque Ndebele. Though today a numerically insignificant people, it is the Ndebele dwellings, guadily decorated, their colorful robes, and their popular beadwork which have brought them fame. Many of the Africans selling beadwork along the Transvaal roads are Ndebele, and more often than not photographs of their homes and clothing are used as illustrations of the highveld scene.

The Transvaal holds many African people whose principal concentration is elsewhere. In the west, the Tswana (also Chuana) overflow from Bechuanaland into the northern Cape and the western Transvaal, their number in South Africa running well over a half million. These Tswana are a fascinating people, often described by South Africans as the most highly developed of all South African Bantu (whatever that may imply), who practice sedentary agriculture as well as stock-raising in poorly endowed parts of the country. In the northern Zoutpansberg Range and beyond are the Venda. These people occupy remote parts of the province, and their distance from the hub of white settlement is reflected by the preservation of their own culture. The northern Transvaal suffers from chronic labor problems, and the Venda will submit to work on the white farms only in time of famine. Otherwise, subsistence agriculture and handicraft work are pursued, and woodwork, metal products, baskets, and pottery are produced, some of which find their way to white collectors. These are the people who do the famed python dance, and they excel in music, possessing a variety of instruments.

The eastern border area of the Transvaal is occupied by the Shangaan people in the north and the Swazi in the south. The Shangaan now number perhaps 400,000 in the province, and their location in the east is the result of their late migration from Moçambique. These are mainly farmers, but they also catch fish in the rivers of the region, which is unusual for Bantu. The fact that the Swazi occupy an area beyond the borders of their country is no surprise. Huge tracts of land were given away by Swazi chiefs in the second half of the 19th century. Today, estimates of the number of Swazi permanently in South Africa fluctuate around 300,000. These people are concentrated in sections to the north and west of Swaziland proper, and although there are considerable differences betwen the course of political events in Swaziland itself and those in the Transvaal, Swazi customs are still similar inside and outside the protectorate. There has been little or no change in the nature of the dress worn by these people, they still treat their hair in a distinctive fashion, and they celebrate each year an elaborate Ceremony of the First Fruits, which is held in Swaziland but attended by Swazi from all parts of the areas they occupy.

There can be no doubt that in general the African occupies the less desirable parts of the Transvaal. Although some land of fair quality has been made available to African farmers, the best land belongs to the whites and is organized in large plantations such as those in the vicinity of Nelspruit. The soil and rainfall maps of the country show one reason for the migration of tribesmen to the cities. The Transvaal includes the heart of the country, the hub of its economic activities, as well as South Africa's most remote parts. The province displays rich farmland and near-desert wastes. A section of its African tribal population is quite advanced in terms of organization and means of survival, while other groups are backward and live a life of precarious subsistence agriculture. More detribalized, "urbanized" Africans live in the Transvaal's great cities than in the cities of any other province of the Republic. This is a land of contrasts, where prosperity and desolation, progress and stagnation are all exemplified.

Durban and Fertile Natal

OPEN ON SUNDAYS

Durban and Johannesburg have a good deal in common, but there are also some very obvious differences between the rushed Golden City and this lazy port town. To many objective observers of the South African scene, Durban is by far the more pleasant city, if only because things move at a less hectic pace. Despite the heat this is quite a gay town, with its colorfully decorated and brightly lit promenade, open-air restaurants, constant stream of tourists, and attractive, wide avenues where cars bearing license plates from all over Africa and even overseas help swell the busy traffic.

Durban's central business district boasts many a skyscraper, and when one enters Durban along the long road from the west, the skyline can be seen to tower over the adjacent suburbs in an impressive manner. Durban in a way is a "one-street town," for the commercial life of the city is dominated by wealthy and spacious West Street, which runs all the way from the beach and promenade through the center of town and on to the Johannesburg road. West Street, like Smith Street and other wide shopping streets in Durban, was designed to allow a wagon and full span of oxen to make a U-turn, and today motorists owe a debt of gratitude to those who laid out this plan. The width and airiness of Durban's streets make the daily heat more bearable. What the narrow stone canyons of Johannesburg's core would be like here in the subtropical heat of the humid Natal coast is a matter for meditation on the part of many an appreciative Durbanite. While Johannesburgers often complain about their city, Durban residents seem to express much more satisfaction with their habitat, and not without reason. The city sprawls over a series of hills, and landward from the beach along West or Smith Street the ground rises to a crest dominated by the odd tower of Howard College of the University of Natal. From this tower and from the beautiful residences along the surrounding slopes can be gained magnificent views of the central city, the Bay, Bluff, and the ocean.

Durban developed on a site once called Port Natal, the bay at the mouth of an insignificant stream. In fact, the old city was originally called Port Natal, the name Durban being derived from the name of Sir Benjamin D'Urban, a British governor of Natal Colony. The bay possesses one outlet, a narrow channel between a sandy ridge called the Bluff and a sandbar extending toward it from the opposite side. The Bluff, which rises well above the ocean and the bay, has become a residential area, and along the sandbar, which is much lower, lie the harbor, the promenade, and the famous Durban beaches. The channel is so narrow that only one ship at a time can negotiate it, and concrete piers have been built to keep it navigable despite accumulating sand. The bay, Durban Bay, is beautiful in its setting. From the city, the Bluff and the harbor form a backdrop to the rather still water on which many sailing enthusiasts practice their sport, and from the Bay itself, the modern skyscraper apartment buildings and tall buildings of the city center make a grand impression.

Unlike Johannesburg, Durban does not die on Sunday, and on some evenings shops stay open a few hours late. On Sunday, a host of city-dwellers join the tourists in driving and walking along the esplanade, and restaurants do a good business. Although theaters are closed by law, there are a number of attractions which remain open to the Sunday visitor, and it would be difficult to prevent fishing along the piers, surfing, and swimming.

To the average South African, Durban and heat are

practically synonymous, and those who do not live in Durban often say they never would wish to live there for this sole reason. It is rather amusing to hear a resident of Komatipoort, in the eastern Transvaal, one of the hottest places in all Southern Africa, exclaim that he could never live in Durban—such a terribly hot place! Yet he has a point, for Durban is humid as well as hot, and Komatipoort is dry. While Johannesburg's coolest month (July) averages about 50°, Durban's coolest month (also July) averages 64°, which is only 3 degrees cooler than Johannesburg's warmest month (January). Durban suffers in the summer. With relative humidity in the 90's every day, temperatures also hover at 90°, the warmest month (February) giving a mean of about 77°. Here Johannesburg scores heavily over Durban. Under these conditions the value of cool nights are recognized, and the low humidity of Johannesburg seems almost priceless to the weary, depressed Durbanites. But fortunately, Durban receives well over 40 inches of rainfall per year, and some cooling is effected in this manner. Durban in June, July, and August is just about ideal, and the plateau residents recognize this fact as they stream to the city by the thousands. They are well received. Hotels which are among the country's finest and beaches which rank among the best in all Africa adjoin each other in a city full of such attractions as a magnificent aquarium, a snake park, a permanent beachside playground, an indoor sports stadium, picturesque cricket grounds, beautiful bowling greens, weekly concerts by the Durban City Orchestra, and numerous movie theaters, nightclubs, and the like. In addition, Durban's citizens and its law-enforcement officers realise the importance of the tourist trade to the city, and their conduct is one of the pleasant surprises of this city.

Durban remains, however, part of South Africa, and it has not escaped the general trend of things. This is not a city consisting only of playgrounds and entertainment, free of the harshness and cruelty of Johannesburg. In some ways, Durban has a double problem, for the city has three major racial divisions in its population instead of two. Not only are there large numbers of whites and Africans, but here also is the largest urbanized Asiatic population on the continent of Africa, a population dominantly Hindu. The Asians have given Durban one of its distinctive aspects. Turning off West Street into Grey Street and several other roads is like leaving a European commercial district and entering an Asian one. Domes and minarets mark the architecture. The Asian has been able to influence Durban in this manner because of accumulated wealth; many Indians are extremely well off, and they have invested extensively in property and buildings. Whatever their political position in South Africa, many Indians have managed to establish themselves as successful storekeepers, owners of service stations, and in many other commercial occupations. As a whole, the Indian population of Durban makes itself felt more strongly than the roughly equally numerous Africans. This, of course, is not surprising, for not only have the Africans been unable to establish themselves as have the Indians in the commercial life of the city, but they would not have been able to stamp Durban in terms of architecture in the way that the Indian has done.

The total population of Durban is about 620,000, of whom about 185,000 are white, about 200,000 are Asiatics, and about 200,000 are Africans, the remainder falling into minority categories, including Colored. This makes it South Africa's third city, only Johannesburg and Cape Town exceeding it in size. Durban is very English. Afrikaans is rarely heard in stores and other public places; it is, in fact, spoken by less than 20 per cent of the white population, practically never by Indians and Africans. To hear Afrikaans spoken in Durban, one might go to the police stations, the post office, and government offices. The national government has been careful to place Afrikaans-speaking public servants in this stronghold of English-speaking South Africa. The importance of maintaining the law of the land in a city as un-Afrikaans as Durban has been recognized by the government as long as it has been in power, and hard work has been done to establish a body of civil servants loyal to the state. Hence Durban has not been allowed to deviate from the policies of segregation. It is only fair to state, however, that in any case there would have been very little actual deviation. Whatever some Natalians say, as a whole the province does not give the impression of being a haven for liberalism, and Durban is no exception. On the other hand, somehow segregation in Durban does not seem as oppressive as it is in Johannesburg and Pretoria, and race relations, although they cannot be described as good, are a little less bad.

Durban's residential areas are divided into those reserved for Africans, those set aside for whites, and those reserved for Indians. Especially in the African areas, slum conditions prevail, but considerable effort is being made to replace the slum dwellings with better, permanent structures. Africans come to Durban for reasons other than mining, for Durban is not a mining city. The harbor affords work for many Africans, and with the recent change from daylight activity only to 24-hour service, much more labor was hired. Many Africans also find work in the large industrial sections of the Durban area, of which nearby Pinetown forms a part. The industrial development of the Durban metropolitan area has been spectacular, to say the very least. Large, modern plants have been and are being constructed, and the Durban-Pinetown area as an industrial region is fast overtaking the Cape Town region, which has long held second place after the Witwatersrand industrial complex. Durban is growing, and the rapidity of its growth is a significant feature of the Natal scene. Based upon commerce and industry, and not upon mines whose remunerative period may be more or less abruptly terminated, Durban thrives. Among the major South African cities, Durban's rate of population growth over the last 25 years is second only to Pretoria. The white population of the town rose from 90,000 in 1938 to 125,000 in 1947, approximately a 40 per cent increase in this period, and to 155,000 in 1957, just under 25 per cent over the last ten years. Recently Port Elizabeth has overtaken Durban in its growth percentage, but over the long-term period Durban's population expansion has been the greater.

As in other cities, the agglomeration of many people in a relatively small urban area gives rise to friction, particularly when living conditions leave much to be desired. In 1949, Africans and Indians killed each other in bitter battles in the streets of Durban, and frequently reports emanate from the police department about vicious fighting in the outlying areas. With monotonous regularity, these reports indicate that police were assaulted and sometimes killed by Africans resisting search parties looking for illicit breweries. Particularly, the African area of Cato Manor often makes the news in this respect. Here it was that in 1959 major rioting broke out as a result of trouble over brewing. Africans can obtain beer at municipal beer halls, but private brewing is prohibited, although it is a lucrative business. The 1959 riots broke out when police intensified their search for illicit breweries, while the people (mainly women) who ran these breweries attempted to intimidate the African customers of the municipal beer halls to prevent them from patronizing these places. American press reports described the illicit beer as "potent enough to propel an automobile," and some descriptions of the riots were much exaggerated. However, the nature of the riots indicated much about the philosophy of people who have lived under dismal conditions for much of their life and whose respect for the law, which, they feel, created their misery, is negligible.

It is perhaps true to say that the riots in Durban in 1959, often described as race riots, were in fact not racially induced disturbances but soon attained significance in this respect as white police, helped by African constables, fought African slum-dwellers. The "shebeen-queens" (female operators of illicit breweries) saw the municipal beer halls as competition and a danger to their high rate of profit from beer which, whatever its potency, was brewed under conditions of poor hygiene. This lack of hygiene and the poisonous nature of the beer formed the justification for police intervention. At first, African patrons of the municipal beer halls resisted efforts on the part of the shebeen-queens to dissuade them from consuming city beer, but when police were observed emptying kegs and destroying illicit breweries, general uproar developed. Thousands of Africans participated, and there was death and injury to both sides. The original quarrel was soon replaced by popular defiance of the police, and Africans who had first been at odds among themselves over the matter of patronizing the city beer halls fought side by side against police.

Subsequent white-African friction in Durban involved the wiping out of an entire party consisting of white and African policemen who ventured into Cato Manor, but violence between Africans and police in Durban at the time of the protest against the Sharpeville shooting in 1960 was surprisingly mild. Although thousands of Africans massed and, blocking the highway from Johannesburg, marched to the city center, hardly a window was broken, and only one burst of shooting, with fatal consequences for four Africans, occurred. When this great march of 1960 took place, Africans were carrying sticks, stones, and other weapons, and the situation seemed extremely dangerous. However, the Africans were persuaded to drop their weapons with surprising ease, and although they continued their march, the mood of the people changed, so that by the time they reached the city center, many were singing and inviting sidewalk watchers to join them.

Ever since 1860, the Indian population of Natal has grown steadily. In that year the first indentured Indian laborers arrived at Durban to be sent to the sugar plantations along the Natal coast. The Africans preferred to remain in reserves rather than work for the whites, and so urgently required labor was imported from India. The progress of the sugar industry of Natal in the period 1860 to 1865 (when the policy of labor exportation was temporarily halted by the authorities in India) is to be attributed to this Indian labor. At the end of their indenture periods, most of the immigrants preferred to stay in Natal, so that new labor was constantly required. Within a few years after the first arrival of Indians in Natal, the Indian population of the territory exceeded that of the whites, and this situation has never changed. In 1870 there were 6,000 Indians, by 1890 over 40,000, of whom more than 6,000 were employed by the sugar industry (compared to less than 3,000 Africans). Between 1890 and 1911 a further 100,000 Indians came and remained in Natal, so that by 1913, when the new Union Government decided to stop the immigration of Asiatics into South Africa, over 140,-000 Indians were permanently settled in the country.

The Natal Indian population today exceeds the white population of the province by approximately 40,000. Within the city of Durban live about 55 per cent of the white as well as about 50 per cent of the Indian people of the territory. Although immigration was stopped, the birth rate of the Indians in South Africa is roughly the same as that of India itself, and this has been sufficient to keep the Asiatic population ahead of the white population, growing by both immigration and natural increase. Religious objections to birth control, early marriages, and financial security of a middle class have created large families, and although the high death rate reduces the gains on the white population somewhat, the increase per thousand of Indians is still double that of the whites. While the white population of Durban is now only slightly smaller than the Indian group, a simple calculation shows that within a half century the Indians will outnumber the whites by two to one and outnumber the Africans perhaps by as much as three to two. Most of the Indians who reached the end of their indenture period became cultivators, and it has taken a long time for them and their fellows to enter commerce and industry. The *Race Relations Journal* published in Johannesburg reports (1947) that over one-third of the working Indian population in Natal in 1936 was engaged in agriculture, only 12 per cent in manufacturing, and 15 per cent in commerce. From the very beginning, the freed Indians made an important contribution to Natal in terms of agriculture: on their small plots, they developed the market gardens which supply fruits and vegetables to the city. Farm incomes, however, have not been high, although some Indian sugar growers and others cultivating fruit and bananas have done quite well.

Some of the Asiatics who came to the city to seek work showed themselves to be astute businessmen, and of late the percentages of Indians employed in industry and commerce have increased. Moreover, among the Indians who came to South Africa after 1860 were some wealthy merchants, who have prospered and who are largely responsible for the Indian aspect of parts of Durban alluded to previously. As propaganda, it can be alleged that opportunity for the Indian is unlimited and that hard work will always achieve wealth, but the obvious wealth of some of the Asiatics in Durban is rarely achieved solely on this basis. On the other hand, an increasing number of Indians in Durban and elsewhere in Natal are being educated, and thus one of the great barriers to their entry into industry is being removed. In 1955, 60,000 white and 76,000 Indian children went to Natal schools. With very limited education, most of the Indians employed in industry find work as semi-skilled laborers. Furthermore, individuals have been successful in service trades such as tailoring, shoe-making, operating service stations, laundries, and the like. The Indian works assiduously, and recognizes the opportunities of Natal as well as its limitations. Since 1911, when some 27,-000 Asiatics returned to India on the basis of a government repatriation scheme, very few Indians have taken advantage of similar schemes. Less than 100 per year accept the financial stipend and free journey to the homeland.

The white people see the wealth of certain of the Indians, their rapid increase in numbers, lack of hygiene, and irreconcilable religious views as a danger to white supremacy. Few thank the Asiatic for helping place Natal on the road to prosperity, and in 1943 the South African Parliament passed legislation to control land purchases, particularly in Durban, by Indians. Indians referred to this as the "Ghetto Act," and it was the beginning of the involvement of the Indian government in the matter of the treatment of Indians in South Africa. Indians, for their part, feel superior to the Africans and stress this superiority in demanding equality with whites. The fact that the Indian is indeed favored is deeply resented by the African, who views with displeasure the economic gains of his competitors. Both the African and the white feel that the Indian is an intruder into Africa. Latent hatred between the two suppressed groups was tragically released in the 1949 riots which, like so many other disturbances, began with a small non-racial incident and grew into three days of uncontrolled racial murder and rape which left hundreds dead and caused thousands of dollars damage. Africans and Indians have not forgotten their differences, though the police enforce peace between them.

"THE BELOVED COUNTRY"

No province in South Africa contains so much beauty per square mile as does Natal. This is Alan Paton's "Beloved Country" at its best, with magnificent beaches, green, rolling hills, stupendous escarpments, unforgettable canyons, and limitless vistas. The western boundary is the summit of the Drakensberg, so that the land of Natal falls in a series of steps to the coast. Large rivers such as the Tugela have carved impressive gorges into the plateau rocks. From the gently rolling cane-covered fields adjacent to the ocean to the high mountains of the west, there is endless variety and unfailing grandeur. In winter, the crest of the Drakensberg and the upper parts of the foothills may be covered with a mantle of snow that can be seen from many miles away, while the viewer may turn around and look down upon palm-lined beaches cooled by the soft sea breeze. Natal in winter is dry, particularly away from the coastal region, and the veld turns gray while awaiting the stir of spring. Harsh cold is rare, though as one ascends the steps leading from the coast to the plateau, even in

Pietermaritzburg, the capital of the province, located just
50 miles inland along the beautiful highway leading on to
Johannesburg, winter may bring chilly nights. The Natal
coast may be uncomfortably warm in summer, but it is ideal
in winter, when it benefits not only from its low elevation
but also from the Moçambique Current, which sweeps
warm water from the tropics along the shores. The prevailing winds do the rest.

Summer brings rain and warmth to the interior, and the
whole region springs to life. The grass turns green again,
rivers renew their efforts, lakes fill up, trees revive. Cattle
which have just barely survived the winter drought return
to health. Now there is work to be done, for the grass
along the roadsides has to be kept short, trimming is
needed once again in the beautiful gardens, and roads damaged by the heavy rains must be repaired. While nature
thrives, man suffers. Now, low elevation brings not the advantages of winter but the troubles of the tropical summer. From Ngwavuma in the north to Port Shepstone in
the south, days and nights are hot, and there is little relief
from the enervating humidity.

Good roads lead from Durban not only to the interior
but also north and south along the coast. The road to the
north leads through rolling hills covered by waving cane,
which grows as far north as Mtubatuba. To the south,
sugar fields occur as far as the very southern tip of the
province at Port Shepstone. As one of the industries invariably associated with Natal, the sugar industry deserves
some attention. The first sugar to be harvested in the province was cut in 1851, and by 1872 Natal's output was 10,-
000 tons. The industry did not develop without difficulty,
for first labor problems (solved by the importation of Indians) and then diseases threatened its survival. The latter
problem was solved by the importation of new cane varie-
ties and by research which has produced sugar canes specially suited to Natal conditions. The rise of the industry
is but one of the many results of the influx of population and
rapid urbanization which followed upon the diamond and

gold discoveries in the Cape and the Transvaal. Railroads were built all along the Natal coast, and the acreage under cane increased constantly. The South African market was for a long time able to absorb all the sugar produced locally, but shipments have been exported regularly since 1953, after the industry had overcome the tremendous setback which resulted from the 1949–50 drought. Presently, the sugar industry provides work for some 70,000 people in Natal, of whom about 60,000 are Africans. The Indians have drifted away from the industry since the early days. Production now exceeds the million-ton mark and is still rising. Irrigation is practiced in some regions where the rainfall is insufficient, and the region into which sugar cultivation may yet expand is circumscribed approximately by the 2500-foot elevation contour: sugar cane seldom survives winter above this elevation. Africans and some Indians own cane land in a few places, but the amount of land they occupy is minimal. Approximately 820 white planters possess just under 90 per cent of the land under sugar cane, while more than 3,000 African and Indian planters have slightly over 10 per cent.

Natal's resource base is, however, not confined to soils and climatic conditions suitable for sugar cultivation. The province possesses sizable coal deposits in the northwest, where lie towns which such names as Newcastle and Dundee. It is from this coal that the electricity supply of much of Natal is generated, and the discovery of the deposits in the 1840's was one of the factors which led to renewed British interest in the territory after the first Boer-British conflict at Port Natal. In the north of the province, irrigation schemes are being tried. In places, extensive afforestations today supplement the small acreage of indigenous forests. Some insignificant mineral deposits are being worked throughout the area. Perhaps most important of all to Natal's economy is the scenic attractiveness of the province, which annually draws tourists to the interior as well as to the coast.

LAND OF THE ZULU

Natal is the land of the Zulu. Traveling inland from Durban and taking the "old road" to Pietermaritzburg instead of the new, four-lane expressway, one passes along the edge of the Valley of a Thousand Hills, one of the world's most beautiful scenic areas and best known for the Zulu and their kraals and huts. Even today, the Zulu are a proud people who resist the subversion of their culture. In Zululand—as distinct from the urban areas which have drawn (and changed) many Zulu—they shun the white man's clothing. Many of their dwellings are still characteristic beehive-shaped structures constructed of wattle bark and branches and thatch, but, as the proper materials become scarce, this kind of hut is being replaced by one constructed of earth and wattle with a conical roof.

The Zulu peoples live throughout Natal and extend far beyond the borders of Zululand proper. In the north they extend into the southeastern Transvaal, some live and work in southern Swaziland, and in the south they reach the border of the province. Everywhere, the Zulu takes pride in his ancestry and is very much aware of his former predominance among the Bantu of the entire subcontinent. Some feel that the severe discipline and alertness of the Zulu of a century and more ago are still reflected today in the group's character. Whatever the reasons, Zulus are excellent workers, and they consider themselves, and are considered by people of other races in South Africa, to be the elite of African racial groups.

The village, or the kraal as it is called, deserves scrutiny. In the days of Chaka and Dingaan, Zulu villages were large, and Dingaan's headquarters at Gingindhlovu (which can be reached by motor road from Durban northward), according to all accounts, was a substantial settlement. Even in the more remote parts of Zululand, however, the breaking up of the ancient tribal organization is evident in the decrease in size of the kraal. Many of the kraals observed in traveling through Zululand today are single-family villages, occupied by a headman, his wives, and children, whereas in the past the great majority of the villages were occupied by the headman, his family, and a number of unrelated persons who placed themselves under his leadership. Whether large or small, the village is circular or horseshoe-shaped, and the dwelling units surround the Zulu's most holy place, the cattle enclosure. Here, the Zulu believes, the spirits of his forefathers linger, and no woman or child may enter it. Often, another enclosure, ringed like the cattle kraal by wattle branches or stones, will be seen to lie outside the village. This is the enclosure for goats and sheep, and no sentiments of holiness are attached to them.

The village is an economic unit, and the whole structure of Zulu social organization is reflected by it. Within the village, the position of each hut is determined by the status of its occupants, each of whom has his or her individual duties to perform. Every single village has its own cattle and its own land for crop cultivation. The division of duties is based very much upon sex. One will never see women herding cattle; this is done by the boys and men, who also do the milking, handling of all milk after milking, and the cleaning and storing of all utensils used in this connection. Women will be seen in the fields, hoeing, weeding, planting. Agriculture is the woman's domain, as is the fetching of water, sometimes over long distances, the making of the famed pots used for this purpose, and the cutting of thatch to be used in hut-building. When Zulu men are seen in the field, they are likely to be carrying long knives for chopping down the brush in preparation for the opening of a piece of land for agriculture. This is as far as their activities on the land go. Once the land has been cleared in this manner, the women take over. There is some cooperative effort in hut-building, for here the women cut the thatch and do the actual thatching, after the men have set up the structure of poles and branches. In all their activities, the women are helped by girls from an early age on, and the men are helped by boys almost as soon as they can do anything at all. The division of duties is an ancient one which left the males free for fighting and hunting. Today, fighting on that scale has been terminated, and hunting is much reduced, so that the males find themselves with a good deal of free time to sit in front of the huts, smoking and conversing.

That the Zulus are creative in several fields is evidenced by their beadwork, with which they decorate themselves liberally, by some excellent wood-carving, and by their fondness for music, dancing, and singing. One of the surprises of nightfall in Zululand is the absence of the incessant drumming heard in so many other places in Africa. Drums are not used extensively, and a real "Zulu drum" does not exist. To produce rhythmic noises, a hide is stretched and beaten with a heavy stick. A few string instruments which produce a wailing sound have been developed, and reed whistles are used also. But while the number and variety of musical instruments used is limited, the Zulu is very fond of singing, and wherever Zulu are working—whether it be women in the fields or building laborers in Durban—they can be expected to lighten their task by joint singing. Some have amazingly beautiful, deep voices.

From the first contact with Europeans, the Zulu were famed for their war dances, and dancing still plays an important role in their everyday life. Some of the earliest observations made of the Zulu hold good today. Despite their pride and their tribal individualism, however, the Zulu are knuckling under to the march of "progress." They are very much entangled in South Africa's problems.

HOPE FOR THE FUTURE

Natal is where the cultures of three continents meet. Of all Africa's regions, this province is particularly complex, problematic, and fascinating. Of the three population groups, none is outnumbered to the extent that minorities are in so many parts of the continent. Each plays its vital role in the economy of the province and, indeed, in that of the whole country. With all its problems and the racial fragmentation of its population, Natal is also the province of moderation, the province which has, in some quarters at least, a conscience. While South Africa's laws are implemented here as elsewhere in the country, the more liberal element in the province, though not as widespread as some would have it, nevertheless asserts itself significantly. Some of South Africa's great leaders have emerged from among these people, and some have become leaders beyond the borders of the land. It was in South Africa, in Natal, where Mahatma Ghandi began to formulate his philosophies of passive resistance. Ghandi was one of the many Indian immigrants to Natal. He arrived in 1888 to remain for twenty years, was active in the Boer War and during a subsequent Zulu rebellion in 1906. Among the Africans, Albert Luthuli, now under virtual house arrest near Stanger, about 50 miles north of Durban, has risen to become one of the most responsible and moderate voices. A Zulu, he spoke before gatherings of whites, Africans, and others in several of the country's centers, and his arrest and detention was one of the consequences of the crisis after Sharpeville. It may be prophesied that Luthuli, who recently received a Nobel Peace Prize, will play an important role in the South African Republic in the near future. Among all leaders and would-be leaders of the Africans in South Africa, Luthuli particularly stands out as a national and not a tribal figure. His moderateness after all that has happened to him is a tribute to his wisdom. Another famed Natalian is Alan Paton, a product of the white community, present leader of the Liberal Party. He is the author of one of the most significant books ever to emerge from South Africa, *Cry, the Beloved Country,* and of numerous other works. Many regard his integrationist political views as so far beyond the limits of the present situation as to be impracticable, but in a country of lopsided extremism, this kind of extremism is a welcome change. Natal, neither the most populous nor the wealthiest of the provinces of South

Africa, holds much hope and promise for the future. Here, the Indian may become a bridge between white and African, even though today there is more evidence of friction than of cooperation. This is where, in everyday life, Apartheid is least strongly felt, and not only the Africans but whites as well as Asiatics voice opposition to the concept. Here is the cradle of the Liberal Party and the recently founded Progressive Party, another opposition party calling on moderates to assert themselves. From Natal may eventually emerge a pattern of interracial cooperation.

| The Cape

As the scene of South Africa's earliest history, the Cape Province keeps some of the traditions of the old colony, while in other respects enormous change has come about. In terms of size, the Cape Province alone covers well over half the territory of the Republic, being somewhat larger than Texas. In population it is second only to the Transvaal, with 5,200,000 people, a little over half of whom are Africans. The Cape of Good Hope has not lost its eminence in the South African scene. Cape Town has become the gateway to the country, and the city's legendary beauty is known throughout the world. As the legislative capital of the country, Cape Town has retained a part of its original administrative functions, which it now shares with Pretoria.

In some ways it is true to say that Johannesburg is the Transvaal and that Durban is Natal. A similar statement concerning Cape Town would be less accurate. While the city is indeed the hub of the province, there are growing towns which compete with the capital city for the wealth of the hinterland. Because of its very size, the Cape Province embraces a very large percentage of the South African coastline. Thus, of the four port cities serving the country, only one lies outside the Cape. In addition to Cape Town, Port Elizabeth and East London serve the interior.

CITY OF TRADITIONS

Cape Town is one of the world's most beautiful cities. Approaching it by sea from Table Bay or by air from the north or east is an unforgettable experience. No description can do justice to its setting. Cape Town lies at the foot of Table Mountain and its two flanks, Lion's Head and Devil's Peak. Table Mountain reaches over 3,500 feet in height, and since it stands rather close to the shore, it limits the area available for urban development. Thus Cape Town is sprawling around both flanks and occupies the slopes of Lion's Head and Devil's Peak. Table Mountain dominates every view of Cape Town. Wherever one may be in the city, the great wall of sandstone is seen to rise high above. It is a view which never fails to impress. Table Mountain changes in appearance with the changing light of day, its moods vary, and with it the aspect of the whole city. At times, condensation of the warm air rising over the mountain creates a cloud which perfectly covers its top; this is the famous "table cloth." A cableway runs to the top of the mountain, and from the sightseeing station some unparalleled views of the city, the Cape Peninsula, and other areas may be had. The top of Table Mountain is flat enough to take walks on. The first signs of the "table cloth," however, make a rapid return to the cableway mandatory. Hikers lost in the mist have plunged off the sheer scarps which bound the mountain.

Cape Town, with over 700,000 people, is South Africa's second city. It is the one large city in Southern Africa with a history that runs into centuries. This historic momentum is evident in the mother city today, and some structures still stand which were there before the Great Trek began or Johannesburg was even spoken of. Capetonians are sometimes accused by other South Africans of feeling and acting superior, of considering themselves the cream of South Africa's population. Whatever effect it may have on its inhabitants, Cape Town does stand alone among South Africa's cities in terms of architecture, tradition, and culture. Environmental determinists might add to this list the fact that the city possesses a mediterranean climate, which singles it out from the rest of the country and indeed from the rest of subsaharan Africa. Efforts have been made to link the climatic environment of the Mediterranean shores to the cultural progress of the Romans, and some would attribute the western Cape's individualism to a similar cause.

Cape Town's summer days are hot, but they are seldom excessively humid, and the nights are almost always cool. January and February, the warmest summer months, average about 70°, though daytime temperatures do rise into the 90's and, infrequently, may tip 100. This is the dry season, for unlike the rest of Southern Africa, this small part of the Cape receives the bulk of its moderate rainfall during the winter. Cape Town's winter can be rather unpleasant, on days when the wind blows hard from the cold south. Still, the coolest month is July and its average temperature is 55°, so that serious discomfort does not last. Cape Town in winter seldom experiences many consecutive days of gray, dull, cold, overcast weather. Skies clear rapidly after a storm, and weeks of sunless days, such as may occur in the middle latitudes elsewhere, are unknown. On the other hand, sensible temperatures, because of the high winds, may be considerably lower than the recorded figures. Foreigners and South Africans from the interior complain that Cape Town homes are not built with due concern for the cold days and that heating them adequately on such occasions is almost impossible.

From Cape Town's point of view, there are negative as well as positive aspects to its lengthy history. Urban development in the central city area has been haphazard, and what has been said of the modern rectangular blocks of Johannesburg, Pretoria, and Durban does not apply to Cape Town. The one modern, wide avenue is the main street, formerly named Heerengracht, then Adderley Street, and now Heerengracht once more, along which the major department stores and other commercial establishments are found. Apart from this thoroughfare, however, there are innumerable narrow lanes, alleys, dead-end streets, and one-way streets which are too congested even for one-way traffic. To disentangle itself, Cape Town is repeating a maneuver successfully carried out first in 1938: the reclamation of a sizable portion of Table Bay for orderly, well-planned expansion of the central business district. This expansion has begun, and impressive, modern multistory buildings are arising on the new Foreshore to give Cape Town a changed aspect.

Cape Town does not as yet possess a core area with the appearance of that of Johannesburg. Although vertical development is in progress, Johannesburg's downtown area dwarfs that of Cape Town entirely. Neither does the mother city possess the peculiarities of Durban—its Asiatic influences, its gay promenade. But Cape Town has its own, distinctive features, many of which are unrivaled in the entire country. If anything, the nearby beaches are superior to those of Natal. There is perhaps nothing in Africa to compare with the drive out of Cape Town toward the west, around the Head and along the Twelve Apostles, a series of imposing mountains cut in sandstone. Beautiful bays, secluded beaches, and marvelously blue ocean water combine to present a bounty of scenery perfect in its beauty. Along the slopes, the suburbs of the city extend, boasting magnificent villas with sweeping views over the Cape's splendor. The old castle, completed in 1676, is still in use today. Government buildings, botanical gardens, and a host of other features, including some excellent museums, help to make Cape Town the Republic's most fascinating city.

In the beginning of each year, special trains arrive from Pretoria with hundreds of government dignitaries and their families. They live in Cape Town for six months while the Parliamentary sessions take place. Cape Town and Pretoria have long been involved in a rivalry over the matter of which will be the country's capital, and so far Cape Town appears to have the upper hand, for it is of course the Parliamentary sessions which make the news. As the administrative capital, Pretoria has its Union Building, but day-to-day interest focuses on the events in Cape Town.

Two racial groups are dominant in the population of the city. There are about 340,000 Colored people, about 290,-000 whites, a mere 55,000 Africans, and perhaps 15,000 Asiatics. The Cape Colored are a complex group, including Malay people, people with white and Hottentot blood, people with white and West African blood, and various other combinations. Slaves, imported from tropical Africa, Moçcambique, Madagascar, and the Far East, mixed with whites as well as with Hottentots. During the first twenty years of white settlement at the Cape, possibly three out of every four children born to slave mothers were fathered by whites. When the pioneer trekkers set out for the interior, there appears to have been some mixture between whites and Hottentots along the frontier. Colored people, however, are not numerous on the highveld, which, considering trekker philosophy, is not surprising. The Colored are even more strongly confined to the Cape Province than the Asiatics are to Natal, and in the Cape they are concentrated in the west.

The Cape Colored people have struggled for advancement over many decades. Their great poverty, particularly during the 1800's, is reflected by descriptions of their dismal living conditions in the Cape's growing towns. Municipal sanitation arrangements often stopped where the Colored "location" began, and it is thought that the Colored suffered as much from the diseases that were rampant at the time as did the Africans, whose plight in this respect is much better known and more often cited. Nevertheless, a small minority of the Colored in the cities managed to improve their lot. These were the skilled and semi-skilled workers, who by their ability and acceptance of lower wages held jobs in almost every trade. After 1880, with the arrival of the Trade Union Movement in Southern Africa, the Colored worker seemed permanently entrenched in all the occupations in which he was active. Before union in 1910, the labor organizations did not discriminate against non-white persons, and some of them even came to have more Colored members than white. By insisting on equal wages for white and non-white, the unions helped to end wage discrimination against the Colored. Unfortunately, however, the fact that the Colored would do work of equal quality for somewhat less money was a major reason for the relatively large number of non-whites employed in skilled and semi-skilled professions; when the unions insisted on equal pay, the Colored, who were supposed to benefit, began to lose jobs to whites. Thus, numerically, the Colored did not benefit as much from the labor movement as was intended. Officially, during the existence of the Cape Colony and prior to Union, there was no official discrimination against Colored workers in government or other work. It was the lack of education which kept their numbers in most jobs very low. It is a tragic fact that while education facilities have greatly improved since union, the opportunities for work in very many instances have been destroyed by government intervention, so that as one avenue of advancement opened for the Colored people, others closed.

Cape Town, at the time of union, was by far the most important source of skilled Colored labor in South Africa. Here, government action with regard to the Colored had its most severe effects. The Juveniles Act of 1921 and the Apprenticeship Act of 1922 both favored the white apprentice. These acts established certain qualifications and facilities for apprentices intending to enter certain professions, and they both discriminated against the Colored. As though this were not enough, the South African "poor white" problem also became most acute during post-union years. Whites rendered destitute by the depression began to stream from the rural areas to the cities. Here, they had to be given work, even unskilled work, and this meant competition for the struggling Colored, who lost more jobs. In the years after 1910, interference on the part of the government and the flow of whites into the cities caused a steady reduction in the percentage of Colored wage-earners and a corresponding decline in the well-being of these people.

Prior to the complete segregation of the universities of South Africa, some Coloreds graduated and went into teaching in Colored schools. Currently, an all-Colored university is being set up in the western Cape, which is supposed to afford these educational opportunities to the Colored once again. There are a few Colored doctors and lawyers, but opportunities for these people remain limited. In Cape Town and in the other towns of Cape Province, the Colored men are almost exclusively artisans, factory workers, or laborers, while many women work as domestic servants or in factories also. They speak the Afrikaans language more than any other, although they mix English and Malay words with this tongue. Most appear to be devoutly religious. For many years, the Colored have sided with the whites in the issues besetting the country, but they have not been rewarded for this. One of the much-publicized actions of the Nationalist government in recent years has been the removal of the Colored voters from the surviving common roll. The Colored feel themselves apart from and above the Africans, and it is of course likely that Colored acceptance of discriminatory policies on the part of the whites stems from the fear of being treated on an equal footing with the Bantu. There are some signs that this attitude is changing. In view of what has happened at the Cape since 1910, it is remarkable that this change is coming so late.

One group of people commonly included in the Colored community of the Cape but in reality quite distinct from the bulk is the Cape Malay population. The Malays, brought originally by the Dutch East India Company from Java, Sumatra, and Malaya resisted intermarriage and remain a rather distinct racial group even today. The Malays are Moslems, and the terms Malay and Moslem at the Cape are virtually synonymous. Actually, the Moslem community at the Cape is much larger than the Malay group, in that Ceylonese, Indians, Pakistani, and even some Africans and whites belong to this faith. A large section of the Cape Malays, like other Colored people, have long used the Afrikaans language as their medium of communication, but another group speaks an Eastern language, with English as a second tongue. Many Malay words are today found in the Afrikaans language, and Malay food is popular with many people, Colored and white, in Cape Town.

Many Moslem residents of Cape Town travel to Mecca as pilgrims, and colorful festivals are held when they depart and return. Of considerable local interest is the Chalifah, a performance with religious significance, in which drum-beaters and other players partake. Designed to emphasize the power of faith by exposing the body to steel knives and needles, the Chalifah unfortunately has been misused for theatrical purposes. In the New Year Colored Carnival held at Cape Town each year, Chalifah performers participate. Most of these performers are very skilled at swordplay and wield fearful weapons. After prolonged drum beating they attain the desired degree of religious ecstasy and then puncture their limbs with these sharp implements.

Like other Colored people in South Africa, the Malays love folksong and other arts. They give the impression of being an unusually cheerful people. There are some good choirs in Cape Town, and there is a Colored ballet company and a Colored opera group. Despite their struggle for survival, their lack of opportunity and lack of hope, notwithstanding the deterioration of their living conditions as exemplified by the Malay section of Cape Town, the Colored place on Cape Town a stamp of individuality which may not be so absolute as that of the Asiatic on Durban but which is no less permanent. They participate in one way or another in almost every aspect of the economic life of the city, and numerically they are more significant in Cape Town than the Asiatics in Durban. Perhaps even more than the Indians, the Cape Colored may be looked upon as one of South Africa's real hopes for the future. Here is a real bridge between black and white. Presently isolated, apathetic, unable to participate in the development of their own country which owes them tremendous debts, discriminated against by the whites and shunned by the Africans, the Colored may nevertheless play an important role in the future of the Republic. The day will come when, if the Colored are not given the opportunity for advancement, they will create it.

Although as many as half the Colored of South Africa are concentrated in and around Cape Town and in the Cape Peninsula, many of them live in villages and on farms throughout Cape Province. It is said that the lot of these folk has not improved materially since the 1840's. In the small fishing villages along the rugged coasts, Colored fishermen eke out a living, but many of those trying to live off the land manage only a very poor subsistence. The poverty of the rural Colored helped cause the flow of these people to the cities. In the Western Cape, where the African population is quite small, farm labor was and is mainly Colored labor. South African whites are not used to paying their farm labor decent wages, so that the Colored are little if any better off than the Africans on the farms of, say, the north central Transvaal. Since the Colored feel somewhat superior to their African countrymen, equal wages hold no appeal. Not much good land is available should a laborer wish to attempt an independent existence as a farmer, and so he treks to the city, whatever the limitations on his employment there.

Frequently, the Colored people at the Cape are accused of drunkenness, which is often cited by South Africans as a justification for segregation. It may be said that in regard to drunkenness white South Africans live in glass houses and that they have contributed directly to the Colored partiality for liquor. On the wine farms of the Cape, it was once frequent practice to pay farm labor with liquor rather than money—the "tot system." To the wine farmer, this was an excellent way of ridding himself of poor-quality surplus wine, and to the Colored laborer a means of forgetting his troubles. The troubles have not disappeared, neither on the farms nor in the cities. The method of forgetting them likewise persists.

The Great North Road leads out of Cape Town, the Gateway to Africa, northeast to Johannesburg and on to the Limpopo River in the northernmost corner of the Transvaal. It is just under a thousand miles from Cape Town to the Golden City and nearly four hundred miles more from there to the Southern Rhodesian border. Almost half the way lies in Cape Province. The Cape seems endless in whatever direction it is traversed. Within its boundaries lie places as far apart as Kimberley and Mossel Bay, Port Nolloth and East London. Along its borders stretch South West Africa, Bechuanaland, the Transvaal, the Orange Free State, Basutoland, and Natal. Its shores are pounded by both the Atlantic and the Indian Ocean. The Cape dwarfs the remaining three provinces, which together cover an area only two-thirds that of the Cape. Somewhat larger than Texas, the Cape Province is no less diversified: magnificent mountains, endless plains, barren deserts, and large rivers are among its scenic features.

The diversification of the Cape is revealed in one look at the map of Southern Africa. While Natal lies between the Drakensberg and the ocean, and the Orange Free State is confined to the highveld plateau and the Transvaal virtually so, Cape Province includes large sections of the interior plateau as well as the bulk of the South African coastline. In the extreme south, the east-west trending ranges present a scenery unique for subsaharan Africa. Between the hills lie fertile valleys with beautiful farms where vegetables and fruits are grown; this is also the country's wine-producing region. The famous "Garden Route," one of the most magnificent scenic drives in the world, passes through these valleys. It runs eastward from Cape Town in the direction of Port Elizabeth and East London.

The Cape's great beauty is not confined to the Cape Ranges, although its character changes abruptly once these parallel chains of ridges are left behind. Between the last of the hills and the Great Escarpment lies a large basin, called the Karroo (a word which appears to have come from the Hottentot *kuru,* which means dry). Actually, as with the terms highveld, middleveld, and lowveld, South Africans do not precisely define the area to which the term Karroo refers. It is geographically correct to apply the name only to this basin separating the Cape ranges from the Great Escarpment, but possibly because of the meaning of the word, much more of the interior of the Cape is popularly termed Karroo. There are some gaps in the scarps leading up to the plateau, and the dryness of the Karroo basin proper is continued above the escarpment. The Karroo is an extensive region which affords some magnificent views but is characterized by sparse vegetation and poor soils. Poets have sung of it, and South Africans appreciate its buttes and mesas, incredible sunsets, and harsh climate. Few places in Southern Africa afford such real wide open spaces. The very special character of the Karroo takes some time to be properly appreciated, but more perhaps than breathtaking ravines and beautiful beaches, this area is impressive in its unchanged grandeur.

The Great Escarpment is not everywhere of the same prominence. In the eastern Cape, opposite Port St. Johns and the entire series of resort and fishing towns, the scarp is sheer and forbidding, separating absolutely the interior (in this case Basutoland) from the coastal regions. Westward from here, the scarp is broken by gaps which are occupied by roads and railroads. Much of the eastern Cape is African land, this being the region of the Pondo and the Xosa. The largest single Bantu area in South Africa, the Transkei, occupies the entire region from the Natal border to near East London and inland as far as Basutoland. Separated from the Transkei by the immediate hinterland of East London is the Ciskei. The remaining Bantu areas of the Cape Province lie in the north between South West Africa, Bechuanaland, and the Transvaal. This is the Kalahari-Kuruman-Taung region, and here live the Rolong, Thlaro, and Thlaping. While the Transkei and Ciskei in the east include some good land, these northern Bantu areas are dry and poor. On the other hand, the Transkei as well as the Ciskei are densely populated, while the northern areas have less population pressure.

The boundary of the Cape with South West Africa is formed by the Orange River, South Africa's greatest river, which is being choked by the encroachment of the desert. The Orange River's drainage basin, along with that of its northern tributary, the Vaal River, dominates the interior of Cape Province. The Orange originates in Basutoland, very close to the shores of the Indian Ocean, and then flows across the southern tip of the continent into the Atlantic. For many miles, the river proceeds slowly through dry territory, and before reaching the mouth, its waters plunge over the Aughrabies Falls, a spectacular gorge of rapids and falls, with the largest single drop about 400 feet. The Aughrabies Falls lie just a few miles to the east of the intersection of the South West African border and the Orange River. For much of its course, the Orange River flows through a rather deep valley, but in some places there is room for irrigation along its banks. The river is of much importance to the entire country, since it affords some excellent dam sites in its upper reaches and provides water to areas which are among the driest in Southern Africa. In the region of its mouth is desert, and due north from Cape Town there is a rapid increase in dryness. At best, most of the interior of the Cape can be classified as steppe, and it is through these regions that the Orange maintains its course.

There are several important cities besides Cape Town in Cape Province. Kimberley, although no longer so eminent as it was in the heyday of diamond-mining, remains a significant town in the north central part of the province. Many mines are still in operation in the Kimberley region, and others which have been worked out evidence the days when Kimberley rose to prominence in the subcontinent. Among these latter is the largest man-made open-pit mine, known as the Kimberley Big Hole, a gaping aperture of tremendous depth dug by human hands. The Big Hole, abandoned and now slowly filling up with water, is a remnant from the time when present mining methods were not yet employed and when thousands of hands were readily available. Today, shafts are sunk adjacent to a diamond-iferous "pipe," and horizontal levels driven into the loose gravel containing the stones. In former days, the pipe was worked from above, simply emptied until it became impossible to go farther. The Big Hole resulted thus. It was excavated to a depth of 1,335 feet, and at the surface it is over 1,500 feet across. An estimated 25 million tons of gravel were removed from the pit until work became impossible because of water and collapsing walls.

Along the coast between Cape Town and Durban flourish Port Elizabeth and East London. Port Elizabeth is the older and larger. It has a total population of well over 200,000, including nearly 100,000 whites. Port Elizabeth competed early in the race for railroad connections with the interior, and it has long been of more than regional importance. Although not a serious threat to the domination of Durban and Cape Town, the city has grown rapidly since it was permanently settled by the British in 1820 and continues to improve its port facilities while attracting more industry. Named after the wife of the first acting governor, the city lies on Algoa Bay, a name given to the coastal inlet much earlier by Portuguese explorers. The city is now the center of the fourth largest industrialized area of the country. There are some twenty shoe factories, and all classes of manufacturing are represented. It is here that the automobile-assembly and tire-manufacturing plants of Ford, General Motors, Firestone, and Good Year have been established.

Port Elizabeth is a progressive, attractive city. Its downtown area, which consists of one main shopping street sandwiched between a scarp and the water's edge, is unable to expand, and as a result a process is taking place which occurs in most American towns: a subsidiary shopping center, which has all the ameneties and none of the congestion of the core of the place, is developing elsewhere. Port Elizabeth's city-planners are proud of the progress that has been made in the general appearance of the town, the eradication of slums, and the development of (segregated) suburbs. Plentiful labor, attractive tax rates for industry, and an abundance of water and power would suggest that Port Elizabeth has a bright future.

East London, also located on the coast, is in reality a river port. The region around the mouth of the Buffalo River was annexed to the Cape Colony as late as 1848, and it was not until the arrival in 1858 of over 2,300 German settlers that the place was firmly established. Names like Berlin and Stutterheim in the immediate hinterland of East London originate from this period. Though in 1910 East London's total population was only about 23,000 and the city has not become one of the Republic's important industrial areas, the potential is there. East London is the center for the wool industry of eastern Cape Province. Wool, mutton, and beef are shipped direct from East London to overseas markets. Citrus fruits and other products from this area also pass through the harbor. Its total population, including some 50,000 whites, is about 125,-000.

Both East London and Port Elizabeth possess small fishing fleets, and in this respect nearby Mossel Bay has also attained some significance. Immediately to the south of the tip of the African continent is a large submarine shelf called Agulhas Bank, and various sections of this shelf form fine fishing grounds. Along the southern Cape, there are a number of small fishing villages which have developed as a result of the close proximity of the Bank, and from Cape Town to beyond East London, all ports have fishing fleets. Trawlers are based in Port Elizabeth and Mossel Bay, and their operators make the bulk of their yearly catch from the areas off the western and southwestern Cape. Surprisingly, the total catch is sufficient to place South Africa among the first ten fishing nations of the world. It is in this western region also that most of the famous rock lobsters originate.

The magnificent forests at Knysna, the serene beauty of the hills at George, the rugged splendor of the escarpment

—all this is the Cape. Nothing in Southern Africa equals the contrasts, the variety, and the breathtaking spaces of this province. Unique in numberless ways, the Cape may have been surpassed by the upstart Transvaal in economic growth and population expansion, but it has not lost its individuality. This individuality derives from its long history, its cultural momentum, and the composition of its population, and it has been unshaken for decades. In many ways, the Cape remains the core of South Africa.

| The Orange Free State Booms

Surrounded by the Cape, Basutoland, Natal, and the Transvaal lies the landlocked Free State, into which the trekking Boers first ventured, across the Orange River and on to the Vaal. This is the real highveld, open, rolling country, dotted with buttes and mesas in some regions and grassy almost everywhere. The Free State is only a fifth the size of Cape Province. It is also the least populous province of the Republic, its total of 1,275,000 people equaling less than half the population of smaller Natal.

As one of the two Boer republics of the second half of the 19th century, the Free State lost its bid for a share of the diamond deposits at Kimberley, and the Republic of the Orange Free State remained over all the years of its existence a country with a pastoral economy. Coal was discovered in the northeast, but the Free State for almost a century had nothing to compete with the diamonds of the Cape, the gold of the Transvaal, or the thriving intensive agriculture of Natal. After union, irrigation projects were carried out, and agricultural production increased continuously, but no spectacular economic developments occurred until very recently.

The Free State, like the remainder of the highveld, was once a land teeming with zebra, wildebeest, and other antelopes. Lions were plentiful and long remained a problem to the cattle-ranchers. Moderately well watered in some but by no means all places, the region occasionally suffered—as it does today—from severe droughts which rendered a pastoral existence even more precarious. The preoccupation of the early settlers with cattle-ranching and the consequent late urbanization of the interior delayed recognition of the agricultural potential of the province.

The north central part of the Orange Free State—the area spreading out due north of Ladybrand on the border with Basutoland—forms the largest part of what is today known as the Maize Triangle. Maize, or corn in American parlance, is the staple food for the African population of South Africa. The triangle enclosing the southern Transvaal and the northern Free State contains the most favorable environment for the growing of this crop, although conditions even there leave something to be desired. What has been achieved in the Maize Triangle on mediocre soils with unreliable rainfall is vitally important to the country.

Corn grows best in an area below 7,000 feet in altitude, where the annual rainfall is between 25 and 45 inches, and where no less than 3 inches of rain fall during December, January, and February (South Africa's summer months). In addition, no dry period should last longer than two weeks, night temperatures must not fall below 60°, and the frost-free period must last at least 200 days. Frosts do occur on the highveld, and it is difficult to find any area of the highveld in any year where records show that all the right conditions were present. All these demands notwithstanding, nearly 60 per cent of the cropland in South Africa is under corn, and most of this land is in the famous Triangle. Experimental stations constantly seek better strains of corn and optimum techniques for sowing and fertilization. Soil deterioration is in part to blame for a yield 50 per cent less per acre than the yield in the U. S. Cornbelt, and rotation systems have been devised to combat it. Pests and diseases plague the farmer, and their effects on yield are obvious. But steady progress is being made in the Triangle, which is the real heartland of the present Republic.

Lying as it does between the fabulous wealth of the Witwatersrand and the ports of the south coast, the Free State at an early stage in its history was crossed by railroads connecting these points. The territory itself benefited little from these arteries. The Cape Town–Johannesburg railroad (De Aar–Kimberley route) even skirted the Free State, but another connection now leads via Bloemfontein.

Bloemfontein was the capital of the old Orange Free State Republic and for some time remained the most important town in the interior. The urban area has a population today of over 140,000, of whom about 65,000 are whites. Throughout the history of the Orange Free State as a political entity, Bloemfontein has retained its position as the focal point of administration and cultural affairs. The city has certain very obvious advantages. It is centrally located not only to the Free State but also to the country as a whole: the railway mileage to Cape Town is only 644, to Durban 427, to Johannesburg 263. Highways of top quality converge on the city, which also possesses a good airport a mere six miles from the city center. As a distribution center for the Republic, Bloemfontein is second to none. The extreme northern border of the country is 625 miles by road, and Cape Town in the extreme southwest is 666 miles.

In climate Bloemfontein is characteristic of the highveld. The warmest month is January, when an average of 74° is recorded. Freezing night temperatures are experienced in June, July, and August. July is the coldest month here, averaging about 47°. About 22 inches of rain are concentrated during the summer months, and the winter is dry and dusty. Winter skies are almost always clear and cloudless, and humidity is very low. In part because of so many cloudless nights in Bloemfontein, the city has become the site for two large observatories. The University of Michigan has established one on nearby Naval Hill, and

Harvard University chose Harvard Hill for the Boyden Observatory, which is maintained jointly by several countries, including the United States, Germany, Belgium, and Sweden.

In recent years, Bloemfontein has experienced an increase in industrial development. One reason for this is the limited capacity of the Vaal River water supply, which is restricting industrial expansion on the Witwatersrand in the southern Transvaal. The city lies near Basutoland and its huge labor reserve and is not far from sources of power. The Bloemfontein area does not rank with the major industrial areas of the country. Dispersion of industries is a prime goal of the South Africans, however, and Bloemfontein will no doubt benefit from these plans.

There is another reason for the recent increase in industrial activity in the city. For many years, the Free State was known only for its cattle, corn, and coal. In almost everything, the province lagged behind. There was no urban development to compare with the three great cities of South Africa. There was no gold or diamond rush. There was isolation of the kind the early trekkers must have dreamed of finding. Even the railroads driven across the veld did little to stimulate rapid development of anything other than coal-mining. Achievements in the corn region were important but not spectacular. Unlike sugar cultivation, the growing of corn did not lead to great wealth for a few; it brought modest success for those farmers who persevered in their efforts. While the Witwatersrand expanded and boomed, the Free State was, in a way, a forgotten part of the Union.

Those who were involved in the mining at Johannesburg and along the Witwatersrand knew that the gold-bearing layers extended from this region downward and southward in the direction of the Free State. It was for long suspected that gold might be found under the grasslands of the Free State, but it was not until fully sixty years after the discovery of gold at Johannesburg that the first strike was made here. Methods of prospecting had reached the stage of reasonably reliable prediction, and on April 16, 1946, the announcement was made that near "Odendaalsrus, Geduld [Bore] Hole 1 . . . has intersected Basal Reef at a depth of 3,922 feet." At this point in history, the Free State ceased to lag behind, and a sequence of development took place which in several aspects eclipsed that of Johannesburg and its surrounding area. In 1951, the first bar of gold was poured at the Free State Gold Mines. From the windswept grasslands arose four mining towns— Welkom, Odendaalsrus, Virginia, and Allanridge—which today vie for attention with Johannesburg. Together, these places, mere hamlets 10 years ago, are occupied today by nearly 175,000 people.

Most impressive of these new towns is Welkom (Welcome). Exceeded in population in the Free State only by Bloemfontein, Welkom's rate of growth is such that the capital's century-old hegemony is endangered. Welkom is called by many the prime example of a well-planned town. It was conceived well before the new gold rush attained present proportions, and most of the usual problems of rapidly growing gold-mining towns were avoided. Even today there is room for an addition of 40 per cent to the present population. Four-lane road arteries pass by modern shopping plazas, and suburbs, industrial areas, parks, and commercial centers have been located in such a manner that they will not have detrimental effects upon one another. Suburbs and mines lie apart, yet close enough so that workers need travel only short distances.

What happened in the Orange Free State is evidence of the great vitality of this country. Shafts were sunk deeper and faster than anywhere else in the world. Production skyrocketed, and unexpected problems were solved with a minimum of delay. Uranium extraction plants were put into operation, and industrial development in the Free State was stimulated as never before. At Virginia, the plant manufacturing acid for use in the uranium-extraction process was reputed to be the largest of its kind outside the United States. Plants processing pyrite (an iron sulphide found in the gold-bearing layers) were built. Coal, explosives, wood, cement, and a multitude of other materials were transported to the new gold fields along newly laid railroads. A swelling stream of labor came to the new fields, where production continued to rise. In less than a decade, this part of the northern Free State was changed from a sleepy corn-and-cattle area to a bustling industrial belt.

Bloemfontein is only 90 miles from the new Free State gold fields, and it is not surprising that this city benefited from these developments. The "new look" of the Free State is also evident in Sasolburg, site of South Africa's big oil-from-coal plant. The possibility of being deprived of its oil supply by some international emergency has long concerned the South African government, and after World War II plans were laid for local oil production. South Africa's geology does not provide oil, but there is much low-grade coal from which oil can be extracted. Actual construction of the world's largest oil-from-coal plant was started in 1952 by SASOL (South African Coal, Oil, and Gas Corporation), and on August 23, 1955, the first oil flowed. The process is American-devised, and American advisers played a leading part in designing and building the Sasol plant.

The plant today furnishes South Africa with about one-seventh of its total requirements of oil. In case of a blockade or interruption of foreign supplies, the country will be able to keep its essential services running. It is also thought that SASOL will stimulate the development of the country's chemical industry. Like ISCOR, which provides domestic steel, SASOL means much psychologically to the country. South African whites want to achieve an ever-greater degree of independence, and SASOL is an important step in that direction.

Whatever inroads economic development has made into sections of the Free State, its political complexion remains much the same. This province is still the Afrikaner's stronghold, and Opposition candidates running for office here always lose. The Free State is the most solid Nationalist bastion of the provinces, even though there are not many voters. The Free State by preventing the entry of Indians made itself the only province where all major racial groups of the country are not represented. The Colored population is concentrated in the south and even there amounts to a mere 15,000. Africans outnumber whites four to one in the Free State, and the Africans are restricted to segregated suburbs, the hostels at the mines, and the laborers' quarters on white farms. The Free State booms, but it has not really changed.

| South West Africa Under the Union Flag

When union came about in 1910, the political framework of Southern Africa attained a measure of stability. The four units of South Africa had been united, a permanent government was established, and it seemed that the road was clear for progress. So much attention had been paid to South Africa, that South West Africa was all but forgotten. South West Africa did not directly enter the Boer War, and the British protectorate of Bechuanaland and the northern Cape effectively separated the Transvaal and Orange Free State Republics from what might have been sympathetic action on the part of the Germans in South West Africa. Moreover, the Germans had their hands full in their protectorate, the battle with the Africans lasting well into the first decade of the twentieth century. Only at the end of 1907 was German rule finally established everywhere in South West Africa.

In August, 1914, when World War I broke out, the young Union automatically sided with Britain, and South West Africa remained part of the German empire. Within South Africa, the conflict brought to the surface a renewed antagonism between Afrikaner and Briton. So shortly after the Boer War, it was not easy for Afrikaners to side with their conquerors, particularly in a battle against people who had been sympathetic with the Boers' cause. As

Afrikaners divided between those who gave priority to the survival and unity of the Union and those who refused to join in a battle against the Germans, actual armed revolt broke out in the Transvaal and the Free State, Afrikaner against Afrikaner. The rebellion was over before the end of 1914, but on both sides men died who had defied the British together. The rebellion was broken by two of South Africa's greatest figures from the Boer War, Generals Botha and Smuts.

When the internal conflict was over, the South African army could turn its attention to South West Africa, where strong German forces remained after the defeat of the Africans. The Union forces, led once again by General Botha, advanced into South West Africa from various points early in 1915 while Lüderitz was occupied and the radio station of the Germans at Swakopmund was silenced by naval bombardment. The ensuing campaign did not last long, and in July, 1915, the German army conceded defeat at Korab, north of the capital, Windhoek. Thus German rule in South West Africa was terminated.

South West Africa did not, however, fall into South Africa's hands directly through this conquest. After the war, the League of Nations entrusted the Union of South Africa officially with a mandate over the former German protectorate. The terms of this mandate have since been a matter for much debate, and South Africa's actions concerning the territory have been questioned. At the end of the war, a number of territories in the world were deemed not sufficiently advanced to achieve immediate self-government. Germany, for instance, had lost not only South West Africa but also Tanganyika, Ruanda Urundi, Kamerun, and other territories in Africa. These territories were individually considered in terms of the mandate best suited to their needs, and countries were appointed as mandate powers. Thus France came to rule Kamerun (Cameroon), Belgium was appointed to govern Ruanda Urundi, Britain took over Tanganyika. On behalf of South Africa, the King of Britain accepted the mandate over

South West Africa. It was also recognized that the needs of each of these territories were not always the same, and types of mandates were classified. Certain territories were classified as "Class C" mandates, and this meant that they were to be governed as integral parts of the appointed power. South West Africa was classified as a Class C mandate and thus came under the direct rule of the Union of South Africa.

With the dissolution of the League of Nations, the Union's grip on South West Africa tightened. The League had specified that mandate territories should be guided toward self government, that the interests of the local people should be protected, that petitions should always be transmitted to the League, and that annual progress reports on events concerning the territories should be made. South Africa has been accused of breaking all these rules under the League and ignoring them after the demise of the League. While Cameroon and Tanganyika are now independent, South West Africa has progressed very little on the road toward sovereignty. In practice the whites living in the territory are represented in the South African government, while the Africans, like Africans in the Republic proper, are regulated but not represented. Boers feel that with the dissolution of the League of Nations, South West Africa came permanently under South African jurisdiction by default. Indeed, little attention has been paid to appeals from various sources for information concerning conditions in the region and for evidence of good will in cooperation with the United Nations.

The United Nations has had no success in attempting to persuade South Africa to renew its mandate over South West Africa in terms of the trusteeship system. In 1946 the United Nations rejected a "petition" by inhabitants of the territory in support of inclusion into the Union, for the validity of such a petition could be readily judged from one look at the 1946 literacy percentage in the territory and the figure of 208,850 African endorsements. Perhaps 1 per cent of these people knew what inclusion would entail, and even this is probably a reckless exaggeration. In 1949, the General Assembly requested a legal opinion from the International Court of Justice on the matter. The Court determined that South Africa must submit annual reports and petitions to the United Nations, that South Africa may not change the international status of the territory without the consent of the United Nations, and (by eight votes to six) that South Africa is not legally required to place the territory under the trusteeship system of the United Nations.

During recent years, increasing objections to the treatment of South West Africa have come from within the territory itself. Most vocal of all are the Herero people, who suffered unspeakable atrocities at the hands of the Germans and who, though not numerous, occupy the northern part of the area. Their champion has long been the Rev. Michael Scott, who has appealed on their behalf at the United Nations and elsewhere. Naturally, the majority of the whites in South West Africa are strongly in favor of South African rule, and it is not easy for the African population to make itself heard. Probably, most of the Africans do not understand even the most obvious implications of the present conflict over their country. The only major uprising to have taken place in South West Africa for decades was that of a section of Hottentots, who objected to certain taxes and who were put down with troops and the bombers of the South African Air Force. This was in 1922, and since then South Africa's hold over the territory has become much stronger.

South Africans, when confronted with the facts of the matter, invariably point to the material progress made in South West Africa since 1919 and seek a justification for retention of the territory on this basis. They then point to the low educational standards of the Africans and see this as further support for the same cause. Indeed, South West Africa is by no means among the poor countries of Africa, and it cannot be denied that South African activity in the territory has in some respects been very beneficial. On the other hand, social conditions are comparable to those in any of the provinces, which is not exactly an asset to South West Africa. There are about 500,000 Africans and 75,000 whites in the territory. The whites dominate as they do in South Africa, but a third element is added to the more than 40,000 Afrikaners and 7,000 English-speaking people. This third group is the German population, numbering over 20,000, mainly descended from the earliest settlers. These Germans give "Süd West" a distinctive flavor, and most of them are successful either in big business or on the ranches. South West Africa, as far as the white sector of the population is concerned, is trilingual, Afrikaans, English, and German all being widely used.

South West Africa has no large cities, and the capital and largest town, Windhoek, has a population of under 40,000, of whom about 18,000 are white. A large percentage of the whites in the territory live in this and the smaller towns such as the ports, Lüderitz and Swakopmund, and the mining towns, such as Tsumeb. Walvis Bay, the largest port, is one of the country's anomalies. It is fully the property of South Africa and is directly administered by the Cape, as part of that province. The remainder of the whites live on some 4,600 ranches and at the widely scattered mines.

In terms of the African population, there are some vivid contrasts as well as some striking similarities to South Africa proper. The degree of urbanization is less, and so perhaps are many of the evils associated with this phenomenon. Africans here outnumber whites by ten to one, not just four to one, but in the towns the ratio is more favorable from the whites' point of view. As in South Africa, there are tribal reserve areas, which cover a relatively small part of the territory. The largest of these reserves are in the extreme north against the border with Angola, where the bulk of the 200,000 Ovambo live. Actually, the entire northern part of the territory is an African reserve area, from the coast to the eastern extremity of another of South West Africa's anomalies, the Caprivi Strip. The Caprivi Strip is the lengthy eastward proruption reaching the Zambezi River, delimited by agreement at Berlin in 1884. Today just another liability resulting from the days of the great rush for land, the Strip is occupied by Africans. It may be a problem area of the future, for it helps divide a rather homogeneous area between no less than five political units: Angola, South West Africa, Bechuanaland, Northern Rhodesia, and Southern Rhodesia. Other northern reserve areas, four in all, are occupied by the Herero. The concentration of Africans at the top of South West Africa has given rise to talk of partitioning. In fact a step in this direction has been taken by the establishment of an administrative dividing line between the northern African areas and the white farm district of the south. In the remote north, there are no mines, no white farms, and little control over the Africans other than through local chiefs. The Africans live a precarious life of subsistence agriculture and herding, and population densities in the limited regions where corn cultivation is possible are extremely high. There is little education here, and even outside contact through missionaries and traders is rare. It is true that these Africans have not been much disturbed by the rule of the Union. Whether this is a stroke of luck for the Africans or another reason to criticize the Republic is a matter for debate.

The isolation and neglect of the northern part of their "mandate" may one day cost the Afrikaners dear. South West Africa covers an enormous area (nearly 318,000 square miles, roughly equivalent to the size of Texas and Louisiana combined), but all the railroads, airfields, and hard-surface highways lie in the area of white settlement. The ports are connected with the interior towns, and they in turn with Cape Province. No rail lines or surfaced roads, however, extend north of Tsumeb. When self-rule becomes a fact in Angola, South Africa's most vulnerable side will be exposed to the wave of African independence. The small total population of South West Africa—

taking into consideration its tremendous size—is understandable when the general aspect of the region is considered. The ever-present escarpment does not, as in Natal,
overlook fertile foothills. Although the territory lies
astride the Tropic of Capricorn, the cold offshore
Benguela Current robs most of the incoming air of
moisture, and the coastal belt is known as one of the driest
deserts of the world. Lüderitz, Swakopmund, and Walvis
Bay, consequently, are dreary, cheerless places. Often a
fog drifts inland without providing any recordable precipitation, and when it is absent, dust storms may be expected.
Only the northeast of the territory gets more than 20
inches of rain, but the soils of the north are sandy. The
escarpment is not everywhere prominent, but there is a
highveld region reaching well over 5,000 feet. This highveld plateau is the backbone of South West Africa, for the
land slopes downward into the Kalahari Basin to the east.
The plateau is a desert area in the south, but in the center
of the territory it provides fair pasture. Most of the white
farms and the capital are located here. In the north, where
the soils are so poor and the water table so high that
cultivation is impossible, the world's largest game reserve,
the Etosha Pan Reserve, has been set aside for the protection of wildlife. Africa's biggest elephants roam these
parts. Throughout South West Africa, it may be said that
soil quality is mediocre at best, rainfall is unreliable, and
where the rainfall increases, malaria and other diseases
occur.

What, then, makes South West Africa the coveted prize
of political bargaining that the South Africans recognize it
to be? Mineral wealth is a prime consideration. The
Orange River, which forms the southern boundary of the
territory, has carried many diamonds from the interior
plateau, where they formed, to the ocean. Currents have
swept them against the beaches, and today these beaches
are exposed and yield millions of dollars' worth of diamonds each year. Although the diamond revenue does not
even remotely approach the returns of South Africa's gold,

it provides an excellent reason for continued South African interest in the territory. The entire southern coastal desert possesses diamondiferous terraces, and further discoveries may be made.

Diamonds account for about 35 per cent of the territory's annual income. Prospecting on the plateau has uncovered other mineral deposits. Tsumeb, the mining town in the north, lies amid ores of copper, zinc, lead and manganese, and recently a large iron deposit was located. Vanadium also occurs in this area. The total value of the annual output of the mines on the plateau is approximately equal to that of the diamond fields. Much prospecting remains to be done, however, and the possibility of other ores being located is by no means ruled out.

Off the shores of South West Africa lives the product for which the territory is perhaps best known—rock lobster. The cold current which sweeps up along the South West African coast may have a detrimental effect upon the coastal climate, but it is attractive to aquatic creatures. Near Walvis Bay and Lüderitz are pilchard grounds, and along many stretches of the coast, from Walvis Bay southward at a depth of between 20 and 25 fathoms, the rock lobsters are caught. Lüderitz is the center in South West Africa for the treatment of the lobsters, which are frozen and canned and sold in many parts of the world. So successful has the industry been, that restrictions have had to be placed upon the annual catch to prevent extinction of the lobster population. The value of the annual lobster and pilchard catches is well over 10 per cent of the total exports, and the fishing industry is growing.

South West Africa is of course not a rich agricultural region. Sedentary agriculture can be carried on in only a few places, notably in the alluvial valleys of some of the streams, and there only with irrigation. Corn yields are always consumed within the territory, and in especially dry years corn must be imported. Hardy Karakul sheep do manage to extract a living from the dry countryside, and their hides and wool amount to about 15 per cent of the value of all exports. There are also many cattle on the ranches, though the total value of beef and butter exported is not great. It is significant, however, that the territory exports at least some agricultural products to the South African market.

That this poorly endowed country depends on South Africa for many commodities is obvious. That South Africa is the "natural" market for the territory's own exports is just as obvious. These factors do not necessarily argue against granting independence, however, because by the same token South West Africa constitutes both a source of materials and a market, not a liability, for the Republic. In a sense, the Republic is exploiting rather than supporting the territory. And compared to several newly formed African nations which subsist on semi-nomadic herding, South West Africa would have a better chance of surviving.

THE CONSEQUENCES

South Africa will not voluntarily relinquish its hold over South West Africa. It has invested heavily in the territory and now claims a return on its capital and skills. It has given the whites representation in Cape Town and Pretoria. It has extended its racial policies into the region and cannot alter them here while maintaining them in the provinces. Most of all, South Africa will go to any lengths to prevent the development of an independent African state near its borders. It fears that the slightest retreat from its position in the United Nations might bring about emancipation of the Africans in the territory and start them on the road to self-government.

By virtue of its domination of the territory, South Africa rules from Cape Town to the Zambezi. The Caprivi Strip virtually pinches off Bechuanaland and, significantly, separates this territory from troubled Angola. The decision at Versailles to hand over South West Africa has put the Republic in control of an area 70 per cent larger than the original Union. Lying adjacent to Angola,

South West Africa is the first South African territory to feel the waxing Wind of Change. With political liabilities such as the Caprivi Strip, the peculiar status of Walvis Bay, a backward and isolated African population, a small but dissatisfied Colored group, and increasingly severe objections from the United Nations, South West Africa may become the scene of considerable friction. Any inability on the part of South Africa to continue its aid—in case of prolonged drought, or of an effective boycott on South African, hence South West African, goods, or of serious difficulties within the provinces proper—could lead to uprisings. So far, South Africa has had all the benefits of holding South West Africa. In time, the territory may present insurmountable problems to Pretoria's rule.

FURTHER READING

CALPIN, G. H. *Indians in South Africa*. Shuter and Shooter: Pietermaritzburg, 1949.

CALVERT, A. F. *South-West Africa*. Laurie, London, 1915.

COLE, M. *South Africa*. Dutton: New York, 1961.

FITZGERALD, W. *Africa*. Methuen: London, 1950.

MARAIS, J. S. *The Cape Coloured People*. London, 1939.

PACKER, J. *Apes and Ivory*. Eyre and Spottiswoode: London, 1953.

PATTERSON, S. *Colour and Culture in South Africa*. Routledge & Kegan Paul, 1953.

STAMP, L. D. *Africa*. Wiley: New York, 1953.

STOKES, C. S. *Sanctuary*. Maskew Miller: Cape Town, 1953.

VEDDER, H. *South West Africa in Early Times*. Oxford University Press, 1938.

WELLINGTON, J. H. *Southern Africa*. 2 vols. Cambridge University Press, 1955.

WELSH, A. (ED.). *Africa South of the Sahara*. Oxford University Press, 1951.

WHITE, A. C. *The Call of the Bushveld*. White Co.: Bloemfontein, 1948.

WOLHUTER, H. *Memories of a Game Ranger*. Central News Agency: Johannesburg, 1948.

Rebirth
of the Republic

On May 31, 1961, a new Republic of South Africa was established. The Union, which had been created on May 31, 1910, had survived for just over a half century. The Union period was a stormy one, culminating in the Afrikaners' return to power, the most extreme application of segregation policies, and the termination of the country's association with the British Commonwealth. It was also a period during which South Africa thrived economically, and in this respect the country assumed the leading position on the African continent.

In many respects, South Africa has emerged from 51 years of Union in a state which resembles that of 1910. There is still the division between Afrikaner and non-Afrikaner, and some of the old bitterness has not faded. The Africans still do not have the franchise. Government exclusively by the white minority continues. The traditional republican view of racial separation has been intensified. Nationalism such as stimulated the formation of the early Orange Free State and Transvaal Republics has likewise gained in strength. Isolationism and chauvinism, religious fervor and fanaticism still mark the mentality of many Afrikaners.

In other ways, South Africa has changed beyond recognition. Great cities which were only towns in 1910 have arisen. A dense network of railroads and roads has been built. All places of some importance are linked by air. Telephone communications exist everywhere. In all the cities and over extensive parts of the rural areas as well, electricity has been supplied. For a country which lagged badly in public service and administration in 1910, South African achievements are remarkable. The national income has risen spectacularly, and dependence on mining has been reduced to the point where both manufacturing and agriculture exceed it in their contribution. There are tremendous afforestations, great dams, large power plants, extensive plantations, famous universities, and excellent hospitals. More money was spent in the Union by whites on the education and health of non-whites than anywhere else in Africa. Great statesmen such as Smuts gave the Union an early reputation of respect and dignity—a reputation which did not survive until May 31, 1961.

The South Africa Act, which created the Union, was written at the 1909 Durban Convention. It was the hope of the founders of the Union that the extremism which prevailed in the Free State and the Transvaal would be diluted by time and contact with the more moderate Cape and Natal. In the Cape, there was a common electoral roll on which whites, Colored, and Africans were placed together. This situation had existed since 1853, and it had been found acceptable to all. At the 1909 Convention, the Cape delegates hoped that this system would be applicable to the Union government, but the representatives from the Transvaal and Free State refused to yield on the issue of white supremacy. The non-whites, for a time, remained on the rolls in the Cape, but nowhere else. Eventually, they were to lose their privilege in the Cape also. The moderation for which many of the delegates at the 1909 Convention hoped did not materialize. The philosophies which prevail in the South African government of 1962 greatly resemble those of the republican delegates who over fifty years ago deliberated the formation of the Union.

Afrikaner Politics

The manner in which the Afrikaners, defeated in the Boer War and reluctantly joined in a Union with their victors, again came to dominate the country must be considered if the present course of events is to be properly appreciated. In the early years of the Union there was some real hope that the issues which divided the whites were disappearing. Many Afrikaners sided against their own countrymen when the question of participation in World War I arose, and there were signs that at least a section of the Afrikaner population saw the Union and a unified nation as an ideal superior to Boer nationalism. The first Union Parliament was comprised of the governing South African Party, led by General Louis Botha, which received Afrikaner as well as English support, and a rather weak Opposition. This Opposition consisted of the Unionist Party, led by Jameson (of Jameson Raid fame), who a decade previous had been set free by the British on the basis of ill health, and the Labor Party. In an effort to keep the South African Party intact, Botha, Smuts, and others frequently spoke of the need for cooperation with the British, hoping thereby to put the English-speaking South Africans who had joined the Afrikaners in this party at ease. The effect was unsatisfactory. Instead of these English-speaking members leaving the South African Party, a growing number of Afrikaners felt that they had no place in a party whose leaders talked of the British Empire in complimentary terms. Among the people whom Botha had collected in his first Cabinet was General Hertzog, and he was among those who felt that Afrikanerdom should come before cooperation in the Union. Hertzog was the most prominent Afrikaner to hold these views, and in 1912 he left the South African Party and the Cabinet and formed the Nationalist Party. Although Hertzog himself later mellowed, the Nationalist Party continued under the leadership of D. F. Malan. In 1948, this became the governing party of the Union.

The South African Party governed South Africa from the formation of the Union until 1924. During these first 14 years, the country faced a number of problems attendant upon industrialization and urbanization, to say nothing of the upheaval precipitated by World War I. Trade unionism developed here as it did in Europe, and serious strikes took place in 1913 and 1914, the strike in the latter year nearly paralyzing the country. Relations between white and African deteriorated considerably during this period, as a result of land shortages for the Africans and discriminatory legislation concerning job opportunities. Botha died in 1919, and Smuts took over as Prime Minister. Some years previous, Smuts, as Minister of Justice, had had leaders of the striking trade unions deported without trial; now, within the first year of his rule, he faced a sit-down movement of Africans in a village near Queenstown in the Cape. Army and police units were sent, and 300 people were killed and injured. Soon afterward, in 1922, there was trouble in newly acquired South West Africa, where Hottentots were subdued with arms and bombers. In the same year there were disruptive incidents closer to the center of the country. Mining companies in Johannesburg threatened to award jobs to Africans which had been reserved for whites, and white miners struck in retaliation. A revolution, now known as the Rand Revolution, broke out, and martial law was proclaimed. Johannesburg, for some days, was actually in the hands of the revolutionaries, but Smuts arrived to personally lead the fight against the movement. Nearly 300 people died before the Witwatersrand returned to normal.

The present and notorious lack of organization among Africans to promote African interests in South Africa has not always been a feature of this country. As early as 1919, an organization called the Industrial and Commercial

Workers' Union (I.C.U.) was formed by an African in Cape Town. Thousands joined this movement, which was aimed at bettering the Africans' wages and at removal of the job-reservation procedures. White leaders watched the success of the leader and founder of I.C.U., Mr. C. Kadalie, with some concern. At the Cape, the Africans and the Colored still possessed the franchise, and there was some reason for cooperation with a movement of this strength. In 1927, Kadalie went to Europe as a leader of the African workers, and it should be noted that as a representative of this kind he preceded more famous African leaders (like Kenya's Dr. Tom Mboya) by some 25 years. However, this and other movements like it have come to nothing in South Africa, whereas they have played an often responsible and constructive role in the politics of many other African countries decades later. It is another indication of what great chances for cooperation have been lost in South Africa. By virtue of the I.C.U. and other less successful movements, the Union might have been the leader of all Africa, an example of progress in social and economic interracial collaboration. Kadalie and his I.C.U. provided the opportunity to learn a lesson which elsewhere in Africa was learned only after World War II. There is evidence that some whites were willing to cooperate; of all people, Hertzog and Malan sent money and friendly messages to the I.C.U. when it was at its fullest strength. But as extremism grew, opportunities of this kind receded into the gloom of the present. Burdened by repressive statutes hastily passed by Parliament, the I.C.U. eventually collapsed, as did other trade unions afterward. Today, there is virtually no organization supporting African workers.

While the I.C.U. occupied a part of the South African stage, political moves brought about a change in the government. In 1924, the South African Party was defeated by a coalition of the Nationalist Party and the Labor Party. The South African Party had absorbed the Unionist Party, so that there were two major parties now, one led by Smuts and dominated by his mainly English South

African Party, the other led by Hertzog and supported mainly by Afrikaners. Hertzog's party won, and he succeeded Smuts as Prime Minister of the country. It was a victory for Afrikanerdom, and Hertzog set to work achieving a greater measure of independence for the country. Largely through his efforts, the Imperial Conferences of 1926 and 1930 were held, at which the status of South Africa was considered. The result was the Statute of Westminster (1931), whereby the Union was accorded the status of Dominion in the Empire, and thus South Africa's complete independence (which is in fact what Dominion status entailed) was placed beyond doubt. There is evidence that this took away much of Hertzog's distrust for the British, the Empire, and the Commonwealth, for he now began to advocate continued close cooperation of Afrikaner and Briton and spoke rather along the lines of Smuts and Botha during the first years of union. His words had similar effects. Again, there was a group of Afrikaners who felt that until complete dissociation from the Crown and things British had been achieved, until Afrikaans was again the one and only recognized language of the land, until the republic was re-established, there could be no such cooperation. Hertzog, thus, was closer than ever to Smuts, but his Nationalist Party split on the issue, just as Hertzog had caused the South African Party to split many years before. D. F. Malan took over what was largely Hertzog's political heritage, under the name of the Purified Nationalist Party. This party became the Opposition party until it achieved victory in 1948. Hertzog maintained himself into the 1930's when, deprived of the following Malan had taken over, he joined Smuts, and the governing United Party was formed.

In 1934, therefore, Smuts, with Hertzog and the United Party, which included most of the English-speaking people in South Africa along with some liberal Afrikaners, defeated the new Nationalist Party under Malan. These were years of crisis, and some of the political shifts which took place must be related to the background of economic stagnation. One of the most crushing blows to progress in South Africa was struck by Smuts at the insistence of Hertzog. In 1936, the Africans were removed from the common voters' roll in the Cape, and they were placed on separate rolls allowing them only to vote into office several white representatives. With that, South Africa lost yet another opportunity to build upon a precious cornerstone of cooperation.

The degree of division which still prevailed among the whites in South Africa was brought to the surface most strongly by the outbreak of World War II. Smuts and the English-speaking people in the United Party saw the inevitability of South Africa's alliance with Britain, but Malan's Nationalists bitterly opposed fighting on the side of the British. Hertzog, caught in the middle, could not decide to join Smuts on this matter, and without his support it was only by the narrow margin of 13 votes that Parliament voted to fight on the Allied side in the war. It was the end for Hertzog in the United Party, and the Afrikaner Party was formed to accommodate those whose feelings were antiwar yet not pro-Nationalist. Various efforts to reunite Malan's and Hertzog's Afrikaners failed against the background of bitter antagonism which had lasted over 5 years. Meanwhile, Malan and many of the Nationalists openly supported the Nazi cause and hoped for German victory. There was actually sabotage of the South African war effort by Afrikaners, involving some of the darkest pages of the history of this country.

The aftereffects in South Africa of the war were more important than the problems which accompanied its beginning. Industrialization and urbanization had been stimulated by the war effort, and racial problems had intensified. The United Party, which with the driving force of Hertzog in the 1930's had been very severe on African interests, did not present a clear-cut policy without him after 1945. There was a great deal of ineptitude in government. There were housing shortages, and the cost of living rose rapidly. Hertzog had died in 1942 and his successor, N. C. Havenga, was better able to cooperate with Malan's Nationalists. It was during this phase of South Africa's history that the doctrine of Apartheid was developed, and with it, a coalition between Malan's Nationalists and Havenga's Afrikaner Party won the 1948 election. For the first time since union, Afrikaners again ruled in South Africa.

South Africa's elections are run along the lines of the British system of constituencies. In 1948, more people actually voted for the combined Opposition than voted for the Nationalist Party, but the arrangement of constituencies happened to be such that more seats in the Senate and the House of Assembly went to the Nationalists than to the Opposition. Thus it is true to say that in 1948 most of the people were against the government. The only people eligible to vote were the whites in the country, who constituted only 21 per cent of the total population. There was no representation at Cape Town or Pretoria for Asiatics, while 9 million Africans were represented by four whites (out of a total of 48) in the Senate and three whites (out of a total of 159) in the House of Assembly. The Colored at the Cape, at this time, were still on the common voters' roll. In addition to the victories of Nationalist Party and the United Party, a few seats were taken in both chambers by the fading Labor Party, and political activity was carried on by some independents who secured two Senate seats. The Communist Party of South Africa got only 774 votes in the Transvaal and 1,009 in the Cape and was not represented in either chamber.

The 1948 election was a turning point in the history of South Africa. Having once again gained the initiative, Afrikaner ideals were now to motivate all government action. The election was fought almost entirely on domestic issues, and the Nationalist Party gained its support on the basis of a clearly defined policy concerning non-whites. Apartheid became the symbol of entrenchment of white domination, and this is what most voters wanted. In fact, it may be said that even among those voters voting for the

United Party, there were many who approved of the segregation policies proposed by the Nationalists but who voted United Party for other reasons, to be found in the long years of Afrikaner-English antagonism. This has become clear through the actions of the United Party in opposition after 1948. On the question which most vitally affects South Africa, Apartheid, the United Party has failed completely to play its role of Opposition and has in fact permitted and encouraged government actions to implement Apartheid. Actually, this could have been predicted well before the Nationalist victory of 1948. Although it never stated its policies in this respect clearly, the record of the United Party prior to 1948—when it was the governing party—includes discriminatory action of practically every conceivable kind. It was during the United Party rule that the Africans lost their voting rights at the Cape, and Smuts and Hertzog combined in achieving this end. Legislation concerning job opportunities, the right of people to assemble and demonstrate, pass-carrying, and a host of other matters was enacted before the Nationalists came to power. Nationalism must be blamed for carrying Apartheid to its extremes, but the Nationalists who gained the day in 1948 were neither the sole authors nor the first to apply the philosophy.

The platform of the Nationalist Party in 1948 acknowledged the "supreme authority and guidance of God in the fate of countries and peoples." Thus victory and Apartheid could conveniently be recognized as the will of God, and the Afrikaners now feel themselves the people chosen to bring civilization to this part of Africa—by means for which they were placed in power in 1948. The Party also stated its intention to foster an even stronger sense of national unity, founded upon a common and undivided loyalty to South Africa. There was to be complete equality of language and culture for Afrikaner and English-speaking South African. What cost the Nationalists a large number of votes was no doubt their statement that a "republican form of government, separated from the

British Crown, is best adapted to the traditions, circumstances, and aspirations of the South African people." However, they took care also to state that there would be no effort to impose this republican form of government upon the people without a mandate from the white electorate to be obtained at a special election. After steadily gaining support since 1948, the Nationalists did indeed hold a special election in 1960, and a majority of voters favored the republican form of government. On May 31, 1961, South Africa became the republic which the Nationalists had promised, and ties with the British Crown were cut.

The 1948 Nationalist platform also promised that the Party would work toward increased mutual confidence and cooperation between the white sectors of the population, but this pledge has been broken. The Afrikaners, who sought to fortify their young language and culture in one of the strongholds of English, always feared that English might encroach upon Afrikaans. Now in power, they sought to prevent this by means which were rather similar to those of Apartheid: schools in which both languages had been taught were changed to either Afrikaans-only or English-only institutions. This, of course, did have the effect of removing Afrikaners and Afrikaans from exposure to English, but it dealt another blow to progress. There were protest meetings from parents as schools, long a place where youth from opposite camps could work and play together, were converted to islands of isolation. It was all to no avail, and the schools, where there was most hope for a decrease in Afrikaner-English tensions, now in many instances intensify these divisions. To the Afrikaner, separation had again been the solution to the problem at hand, and the consequences may temporarily have seemed satisfactory, but they will be regretted.

The most significant section of the 1948 Nationalist platform dealt with separation of the races. The Party believed the Christian guardianship of the white races over the non-white races to be the basic principle of its policy.

The non-white races were to be granted the opportunity to develop themselves in their own areas, "in accordance with their natural talents and capacity." "Miscegenation" was opposed, and the territorial, residential, and, so far as possible, economic separation of the races was favored. This, in brief, is the policy of Apartheid for which the Nationalists and South Africa have become infamous.

The Afrikaner Party, which shared the views of, and acted in coalition with, the Nationalists in 1948, but it was not until 1952 that the inevitable merger of these two Afrikaner groups took place. Thus, after many years of division, Afrikaner elements were unified. There have since been small splinter groups with dissident views, but the bulk of Afrikanerdom now presents a single front in the politics of South Africa.

The United Party platform contained no counterweight of liberalism to oppose the Nationalist doctrine of Apartheid. The strongest statement concerning racial relations was that the party would "maintain and vindicate white leadership." But the party also promised to promote democracy and strengthen ties with the Commonwealth, apparently unconcerned about the fact that the maintenance of white leadership and the promotion of democracy might not be compatible processes. Unity, mutual respect, and the development of a true South African culture were called for, as well as the preservation of free enterprise, stimulation of economic development, and the creation of housing and education opportunities for all. The party did promise to uphold the rights of trade unions, but there was no clear-cut statement outlining the manner in which this and other promises were to be carried out. The Nationalists were most explicit in their intentions, and their victory was no surprise to those who knew the mood of South Africa in those days. On the basis of its platform, the failure of the United Party to become an effective and responsible Opposition was equally predictable. It has been a tragic fact in South African politics that the zenith of Afrikaner power coincided with the lowest ebb of non-Afrikaner opposition.

This has led to excesses which could, by energetic political activity, have been reduced in extremity if not avoided.

Having gained the initiative, the Nationalist Party now set about the task of increasing its strength. Within months after the election, Parliament passed the South West Africa Amendment Act. By this Act, South West Africa's white residents obtained representation in both houses of Parliament: four members in the Senate and six in the Assembly. The government was fully aware that the white population of the mandate territory was dominantly Afrikaans and German and that the elected members would be Nationalists. In the special 1950 elections, this is precisely what happened, and the Nationalists had achieved a dual aim: the party majority had been increased, and South West Africa was drawn more tightly into the framework of the Union. In 1952, the merger of the Afrikaner Party with the Nationalists did not actually alter the power of the Afrikaners in Parliament, but it did promote unity in the ranks.

The leaders of the Nationalist government were all very competent men in 1948 as they are today, and it is a dangerous error in the interpretation of South African events to assume that Malan, the new Prime Minister, and his Cabinet were blind and uncalculating in their actions. Accompanying the eclipse of the United Party, on the other hand, was a complete absence of sound and competent leadership. General Smuts died on September 11, 1950, and there was no one with his stature or competence to take over the responsibilities of party leadership. J. G. N. Strauss was elected as the new leader, but he failed to infuse the party with the required stimulus for renewed and vigorous efforts. Malan, meanwhile, rapidly gained in stature as his policies began to take effect. Malan was surrounded by the most capable people in the country (disregarding their beliefs), each of whom was to play a vital role in the ensuing decade. T. E. Dönges was party leader in the Cape, and as Minister of External Affairs he traveled to New York to defend the South African case before the United Nations. Though not an exciting speaker, Dönges possessed a brilliant mind, and his arguments in Parliament contained some of the most lucid analyses of the course of events in South Africa that can be found anywhere. J. G. Strijdom was the Transvaal National Party boss and a true representative of Transvaal Afrikaner opinion. Square-jawed, unyielding, and a tireless worker, he was to succeed Malan as South Africa's second modern Nationalist premier. In Natal, N. C. Havenga, who was one of the most able Ministers of Finance the country ever had, led an unsuccessful Nationalist Party. Havenga was given the job of heading the Natal Nationalist Party after the merger of his Afrikaner Party with the Nationalists. It was a sound political maneuver; Havenga had been a Hertzogite who, although opposing South Africa's joining Britain in the war against Germany, had refused to join the extremist Nationalists of 1939. Of all Nationalists, Havenga was most acceptable to Natal, where he was seen as a moderate Nationalist, particularly as Malan's actions began to take effect. There subsequently was a brief flurry of enmity between Malan and Havenga and a possibility that Havenga would take moderate Nationalists with him in a new party, but this did not materialize. In the Orange Free State, C. R. Swart led the party, and it is he who is today the President of the new South African Republic. Thus the Nationalist Party gathered a group of competent senior members, all of whom were experienced in leadership and the promotion of well-defined Afrikaner ideals. During the following years, they found themselves in important positions of varying kinds, and, unlike their competitors in the United Party, they displayed a high degree of enthusiasm, drive, and ability.

THE WORLD WATCHES

It was during these first years of Nationalist government that world interest began to focus upon South African affairs. Even previous to the 1948 election, South Africa had been involved in a dispute with the United Nations

over South West Africa. When, after 1948, Malan's policies for South Africa were also applied to South West
Africa and the Union ceased to cooperate with the
United Nations, the country came to be the center of world
attention—hostile, critical attention which drove the
country ever deeper into isolation. The treatment of Indians in South Africa was protested before the United
Nations by the government of India, and in 1949 the Durban riots between Indians and Africans took place. In Parliament itself, Malan directed his attention to the Colored
voters remaining on the common roll. When, in 1936,
Hertzog and Smuts collaborated in the removal of the
Africans at the Cape from the common roll, Malan had
argued that the same justifications existing for this action
applied also to the Colored people. Hertzog, however, had
then opposed Malan, stating that he felt that Colored
people who had adopted the white man's culture and language should be included politically, industrially, and
economically with the whites. Malan's amendment aimed
against the Colored failed. Now in power, he renewed his
attack upon the Colored, who had remained on the common
roll throughout the history of the Union, surviving several attempts at their removal. It was a declared intention
of the Nationalist Party to disenfranchise the Colored, and
to this end, the Separate Representation of Voters Act was
introduced into Parliament. Each house passed the bill, but
four Colored voters took the matter to the Cape Provincial Division of the Supreme Court and subsequently to the
Appeal Court when their first plea failed. The Appeal
Court ruled the Separate Representation of Voters Act
invalid, and it seemed that Malan's attempts against the
Colored had finally ended in failure. Naturally, much publicity was given to the entire matter, not only nationally but
internationally, and had Malan and the Nationalists
abided by the Appeal Court's decision, the whole matter
might well have brought some credit to the country after
all. Unfortunately, the decision was actually a reversal of a

stand taken by a similar court in 1937, when Africans, de-prived of their votes, tested the Act in a similar manner and lost. Malan declared that chaos would prevail in a country where successive Courts of Appeal could reverse decisions as important as this and stated that it was "im-perative that the legislative sovereignty of Parliament should be placed beyond any doubt so as to ensure order and certainty." With this, an entire new phase began in South African politics. The Afrikaners, ruling as they were, found themselves stifled by a Court which might pre-vent the attainment of their ideals. They had waited for forty years to realize their ambitions, and they were not now prepared to allow anything to stand in their path.

In April, 1952, Parliament established a High Court which could review any decisions made by the Appeal Court concerning legislation passed after 1931. This High Court, not surprisingly, reversed the Appeal Court decision concerning the Separate Representation of Voters Act. The Supreme Court then ruled the High Court itself invalid, calling it, not a court at all, but just a "parliament function-ing under another name."

The party next attempted to amend the South Africa Act of 1909 itself. This Act determined that there should be continued representation for non-whites in the Cape, where it had existed since 1853, and that any amendment to the Act should occur only with approval of a two-thirds majority in Parliament. The Nationalists thus began trying to line up a two-thirds majority by seeking support among the conservative United Party members. In the 1953 elec-tion, they worked feverishly to gain as many seats as pos-sible, and they made substantial progress. In both cham-bers, they unseated several United Party members. After the election, the Nationalists submitted the South Africa Act Amendment Bill to the joint sitting of Senate and As-sembly, but they failed to obtain the required two-thirds majority. Another bill was introduced, this time not in an effort to amend the South Africa Act but to validate the

Separate Representation of Voters Act. Again the two-thirds majority could not quite be mustered, the bill failing by only 9 votes.

All this widely publicized political bargaining to remove a few Colored voters from the common roll did a tremendous amount of damage to South African prestige everywhere. While so much else might have been done in South Africa, minds and resources were being spent in this costly battle, and it was rapidly driving an ever deeper wedge between Government and Opposition, white and non-white, Afrikaner and non-Afrikaner. Malan retired as Prime Minister and leader of the Nationalist Party on November 30, 1954, having failed to achieve one of his major objectives, having split the country, and having done serious injury to its name. True, Malan had worked tirelessly to implement his election campaign promises as far as they concerned non-white policy. Great strides had been taken in effecting the racial separation as defined by the philosophy of Apartheid. Even more than before 1948, segregation was now strictly enforced in buses, public conveniences, parks, elevators, football stadiums, and other facilities. Yet when Malan retired, South Africa still had a long way to go to reach its present state of Apartheid. In time, while his successors J. G. Strijdom and H. F. Verwoerd were in power, Malan in South Africa actually came to be called a rather moderate Nationalist.

J. G. Strijdom was a Transvaler, and he held all the extreme views of the north. When he took office in South Africa, there was a wave of despair among non-Afrikaners. He had long been known for his extreme views, and it was feared that he would go too far too fast in the application of segregationist policies. Once in office, however, he appeared somewhat less fanatical than he seemed before 1954. There was no relenting on the issue of the Colored voters, though, and in June, 1955, an Act aimed at the reconstitution of the Senate (the now infamous Senate Act) was submitted to each house and passed by both. Since the two-thirds majority in joint session had not been obtained by the 1953 election, this Act was intended to increase the total number of seats in the Senate in such a manner that the balance would swing more decidely in the Nationalist's favor. There would be 89 seats in the Senate instead of 48, and provincial representation, rather than being equal, was now determined by the number of white voters in each province. This meant that in the Cape, Transvaal, and Orange Free State, all newly added Senators were Nationalists, while all those from Natal were United Party members. This meant, of course, that in the event of a joint session there would be well in excess of a two-thirds majority of government supporters. The Prime Minster did not make any effort to disguise the reasons for this action, and neither could he have. He declared that the reconstituted Senate now insured the sovereignty of Parliament and that the separate representation of Colored people was assured. The leader of the Opposition, Mr. Strauss, declared that his party had decided to test the validity of the Senate Act in the courts, but since the Appeal Court had itself been reconstituted in a manner similar to that involving the Senate, the Colored voters finally lost their joint voting rights.

While the Nationalist Party was involved in this and other battles with the United Party, the latter suffered from disunity, particularly after the 1953 election debacle. A small group splintered off, calling itself the Independents and professing general agreement with the United Party platform but voting independently from that party on several occasions. This further weakened opposition to the Nationalist cause. Also in 1953, the Liberal Party was founded. This party stands for integration and direct representation *for* African and other non-white voters *by* Africans and other non-whites. This is the one party in South Africa which opposes racial segregation, but it has no more than a few thousand white supporters, a significant reflection on prevalent opinion in the country. Its present leader is the embattled author Alan Paton, who has one of the most publicized passports in history.

Fragmentation and division continued to mark the Opposition throughout the post-1948 period. The Federal
Party favors greater autonomy for the provinces; with the
coming of the Republic it ceased to have what little significance it possessed. This party was a Natal product resulting from fear of Afrikaner domination of this Englishman's stronghold, and it has never had an election success.
An important feature of the political scene after 1953
was the total defeat of the Labor Party. In a way, the Labor Party's few representatives in both houses were there
only by the grace of the United Party, which declined to
contest constituencies where a Labor candidate was cam-
paigning, in order to avoid splitting the anti-Nationalist
vote. This was done by agreement in 1948 and 1953, but
there was no such agreement in 1958, and Labor was no
longer represented. Among the latest political organizations is the Progressive Party, which initially seemed to favor outright multiracial franchise based upon certain qualifications, perhaps along the lines of Southern Rhodesia
(which in South Africa would actually be a step forward!), but these plans were shelved in an effort to attract
more voters. The one lesson that must be learned from the
voting records and campaign platforms in South Africa
since 1948 is that well over 95 per cent of all the white
people are in favor of segregation. In case anyone had
doubts about this prior to the 1958 election, the leader of
the United Party stated unequivocally that the United Party
was not an integration party, had never been an integration
party (there is ample evidence for this), and would never
be an integration party. It is one of the anomalies of the
South African scene that the Opposition should be credited
in the eyes of the outside world and even in the eyes of
some misguided South Africans with being an anti-segre-
gation party. Indeed, it does favor some moderation, but it
has few scruples about racial discrimination. To oppose
segregation is political suicide in South Africa, and on
matters involving white supremacy, the United Party is the
most loyal of Loyal Oppositions. If people in South Af-

rica's United Party who incessantly claim that they oppose segregation base their choice of party on this matter, then they are all in the wrong party and should join the Liberals. What better evidence is there of South African attitudes than the fact that when the Progressive Party wished to appeal to *moderates* in Natal and elsewhere, strategists considered it imperative that all talk of multiracial voter rolls be shelved until some support had been gained by other means?

In the 1958 election, the Nationalists again increased their majority over the combined Opposition. In the ten years of their tenure, notwithstanding all the political strife, they had accomplished much. Economically the country was stronger than ever. While elsewhere African nationalism was gaining the upper hand in the struggle for independence, South Africa had faced only minor problems in this respect. There was passive resistance on the part of Africans and Indians in 1952, but it was dealt with by special legislation, and it collapsed. While Kenya was undergoing the terror of Mau Mau and violent opposition arose against Federation in the Rhodesias and Nyasaland, South Africa experienced nothing of the kind. A Suppression of Communism Act was passed, and the Russian Embassy at Pretoria was closed, its staff ushered out of the country. Universities which for decades had opened their doors to people of all races were segregated. Politically undesirable persons were exiled, African leaders suspected of agitation were jailed. It may be said that the white man in South Africa never had it so good as he had it after 1948, and the Nationalists used this political bargaining point to great advantage.

Although enemies of the Nationalist regime were constantly forecasting the doom of the country, South Africa continued to prosper. Cities grew, the building boom was virtually continuous, the national income grew ever faster, immigration, though never spectacular, increased. Even African salaries partook in this upswing. A serious attack was made on the housing shortage, and hospitals, schools, and other facilities were constructed. The face of South Africa was changing as never before.

There were other changes, less pleasant than the foregoing. Several Acts were passed between 1948 and 1958 which allowed the Nationalists to alter dramatically the prevailing condition of ill-defined racial separation to one of legal and strict segregation. Most far-reaching in this respect was the famous Group Areas Act, which was the result of long planning and a first giant step in the implementation of Apartheid. The passage of this Act allowed the government to begin a division of all residential land in all South African cities into sectors, each for one racial group. It became illegal for a person of a certain race to purchase land in a residential area set aside for another racial group. There were enforced removals of whole suburban populations to new areas, and there was hardship, something to which the African has become accustomed since 1948. The division of the cities into racial residential areas was only one step in the fragmentation of South Africa on a racial regional basis, such as is now in progress. Another Act, passed soon after the Nationalist takeover, was the Bantu Authorities Act. As elsewhere in Africa, as a result of the urbanization and industrialization of the country, tribalism had been decaying for years in South Africa. But the Nationalists were aware of the value of loyal and powerful chiefs, and they moved to reinforce tribal systems. This process was facilitated by the Bantu Authorities Act.

One of the Acts which caused most resentment was the Bantu Education Act, which removed control over the education of Africans from the provincial education departments and transferred it to the Union Department of Native Affairs. It was considered necessary that there should be uniformity in these matters, and that teaching personnel, materials, and funds should be controlled by the central government. Support for mission schools was cut, and the control of government over African education became ever stronger. It is true that South Africa pays much more for the education of her Africans than do other countries in Africa, but it is a matter of self-interest in that control of education means a large measure of security against subversion. The driving force behind these and other Acts was energetic and able H. F. Verwoerd, who became the Union's last premier after Strijdom's death.

One of the notorious measures taken by the government is the Immorality Act, which along with the Mixed Marriages Act was directed against sexual intercourse and marriage between people of different race. South Africa is a country that defines immorality as miscegenation between people of dissimilar racial groups, and in the South African press brief notices appear with monotonous regularity of people sentenced to prison for having been found guilty of this crime. One of the problems involved in this is that one has to know what racial group one belongs to, and in order to classify everyone, the Population Registration Act was passed by Parliament. As a result of this legislation, every South African resident must carry an identification card which indicates, among other things, his race. This classification procedure has given rise to some dramatic discoveries of people who had for decades passed as white and who suddenly found themselves classed as Colored, or Colored people who were classified as African. The impact of such an event is beyond description, for a family which has passed for white and is then classified as Colored must move to another neighborhood, must take its children out of the white school and send them to a Colored school, and most probably the breadwinner must leave his job, which is likely to be reserved for real whites. While such events have fortunately not been common, the stories connected with those that have occurred are hopelessly tragic, and the pointless efforts of the families to have the classification repealed make some of the saddest tales of human suffering ever told.

In 1958, the Nationalist Party was able to point to the material progress made in South Africa and the vastly increased control over millions of Africans and other non-

190 | AFRICA SOUTH

whites. The party had taken strong action against some 156 of its opponents—agitators, suspected Communists, and others—who in 1956 were arrested and became the subject of the now famous Treason Trial. Opposition to racialism in South Africa is all too easily defined as treason, and these people faced severe sentences. However, although it seemed in 1957 that they would be sentenced rapidly and conveniently, the prosecution fumbled, the defense was brilliant, and the judges honest, so that after years of legal argument which rendered many of the suspects bankrupt, charges were dropped, and most were set free. But in 1958, the case was just under way, there was widespread publication of the charges of subversion, Communism, and other illegal activities, and it appeared that the government had the matter firmly in hand. It all amounted to a record which appealed to South Africa's white electorate, and the 1958 election was another triumph for South African Nationalism.

AFRIKANERDOM COMPLETE

In the same year, Strijdom fell ill, and he died in August, leaving the Nationalist Party to find itself a new leader and Prime Minister. To the surprise of many, and apparently after some heated argument and several ballots, H. F. Verwoerd, the former Minister of Native Affairs, emerged victorious. Like Strijdom, Verwoerd was a product of the Transvaal, and he was an even more absolute segregationist than Strijdom had been. He abolished all representation for non-whites in the Parliament, where until then non-white interests had been represented by a handful of white members. He also promised an acceleration in the implementation of regional Apartheid plans, which he helped formulate in earlier years. Soon, he was making speeches outlining his schemes, and in traveling through the country one could see on bulletin boards in the towns and cities maps of the Union with the country divided into a number of segments, some for whites, others for Africans only. So complete was the mandate given by the electorate to the Nationalists in 1958 that even Afrikaner opposition to these proposed schemes could be all but ignored.

Verwoerd also held strong republican views, and early in his term he declared that he would work toward more independence for South Africa. This was a matter of which most Afrikaners heartily approved, and it formed a good counterweight for the inevitable opposition—even from loyal Nationalists—that was to be expected with the execution of regional Apartheid plans. Meanwhile, world criticism of South Africa's racial policies mounted, and South African goods were boycotted in places. Almost universally, an explosion was expected. The Johannesburg Stock Market began to display an alarming degree of irregularity, immigration figures dropped, investment figures dwindled, and even in usually optimistic circles in South Africa there was concern. The Treason Trial dragged on, becoming more and more a liability, university professors left the country, and signs of coming trouble were omnipresent.

Trouble occurred early in 1960 when a small, well armed, and nervous police party stationed at Sharpeville found itself surrounded by thousands of Africans who had assembled in what was, until the moment of shooting, a non-violent demonstration against the pass laws. This initial demonstration against the pass laws was organized by the Pan Africanist Congress, a group of Africans working for the advancement of their people, particularly in the Transvaal. The police party at Sharpeville opened fire, and before twenty seconds had passed, dozens of Africans were lying dead, many others were injured, and Black Week had begun. About 20,000 Africans marched peacefully from Cape Town's largest African township to the center of the city, led by a student who demanded to see a Minister of the government. This student promised to send the crowd home if his wish was granted, and it was. He sent thousands of people home peacefully with a few words—and was shortly thereafter arrested. In Durban, thousands of Africans marched to the center of the city, actually blocking the main road from Johannesburg. In Johannesburg there were strikes, and there were fights between strikers and Africans who refused, or did not dare, to strike. There was violence in many parts of the country, though it was sporadic. The Verwoerd government, facing its first crisis, called out the Army and Air Force and proclaimed a state of emergency over most of the country. The trouble was soon quelled, and the country returned to normal, such as that is in South Africa.

The ineffectuality of the Africans' 1960 uprising indicates the complete lack of adequate political organization of the non-whites in the country. The opposition which led to Sharpeville and martial law was organized by the Pan Africanist Congress, a militant body centered in Johannesburg. This group which was preceded by various others, jelled into its present state as late as 1959. Led by Mr. Sobukwe, it gained rapid fame when it played a prominent role during Black Week and was the subject of police investigation soon afterward. In deciding on the nature of the opposition to the pass laws prior to Sharpeville, the leaders of the Pan Africanist Congress had not been able to reach agreement with members representing the African National Congress, which centers on Durban. The African National Congress was founded in 1912 and has as its major objective the extension of democratic rights in South Africa to the African people and the advancement of African people generally. Its present leader is A. Luthuli, most respected of the African political leaders in the Republic. Moderation has been the hallmark of the A.N.C., and as a result it has not become the undisputed leader of African resistance to white supremacy. It has, however, outlasted other organizations with similar aims, partly because of this very moderation. The last time the A.N.C. was active as leader of a resistance campaign was in 1952, when it collaborated with the Indian Congress in organizing a large-scale breaking of discriminatory laws. Thousands of Africans were arrested and jailed during this program of passive resistance, and the government passed Acts giving itself greater power to deal with such activities. Since then, the A.N.C. has done little to further its cause.

Under the present circumstances, it is not surprising that passive resistance does not appeal to Africans in South Africa and that the A.N.C. no longer holds the position of the African's leader. There have been various splits in the Congress during its existence. In 1943, excessive moderation was the issue on which one group broke away from the Congress to form the African Democratic Party, mainly confined to the Transvaal. The African Democratic Party professed to oppose Communism, as well as the A.N.C., as being too cooperative with the white rulers of the country. The A.D.P. was one of the several bodies that preceded the formation of the Pan Africanist Congress.

The African National Congress, along with other nonwhite political organizations, was severely harassed by the police after the Black Week of 1960. Leaders were arrested, documents impounded, and several organizations were declared illegal. When it was last possible to obtain some data concerning the Congress, there were about 25,-000 members, but Chief Luthuli, having publicly burned his pass in support of the anti-pass law activities, was under arrest. Luthuli is one of the few non-white figures who have succeeded, in South Africa, in becoming more than local, tribal leaders. He has spoken before mixed audiences in such places as Pretoria and Johannesburg, prior to police-imposed restrictions on his movements. He always asked for moderation and cooperation, for amelioration of extremist views on either side, and, above all, for non-violent action. Luthuli seemed to be the personification of South Africa's one hope for the future on such occasions, and though there were disgraceful efforts to break up some of his meetings (particularly the activities of a group of students of the University of Pretoria) he always aroused much enthusiasm. But even in the 1950's, he was too late. Africans spoke of him as a traitor, one who would sell out to the whites. The whites saw him as a danger, one who would arouse African nationalism as other leaders had done elsewhere on the continent. He was forbidden to speak, then placed under arrest. With his removal from the national scene, another bit of hope was destroyed.

In 1920, the South African Indian Congress was created, to combine and coordinate the activities of the old Natal Indian Congress, which had been formed by Ghandi in 1894, with those of the Transvaal and Cape Indian Congresses. Headquartered, of course, in Natal, and always somewhat better financed than the African political organizations, the Indian Congress has nevertheless done little to affect the country's political scene. Although it has drafted petitions and sent representatives to appeal to India and the United Nations for assistance in the battle for Indian political rights, the Congress has been most ineffective. Several members of the Congress were involved in the Treason Trial, and the Congress itself after Black Week suffered a fate similar to that of the A.N.C. Many prominent men were arrested and held without trial.

Among the other all-white, non-white, or multiracial organizations that have seen the light of day in South Africa or exist at present are the now-defunct Industrial and Commercial Workers' Union (I.C.U.), the All Africa Convention of 1935, which saw non-white unity as its prime aim, the all-white Congress of Democrats, which was declared illegal after Black Week, the S.A. Congress of Trade Unions, and the South African Colored People's Organization. All these have played some small role in the political arena, but none has acquired real significance. It is a tragedy that the Africans and other non-whites, who should be united in the face of their present common enemy, are so hopelessly divided.

Whenever the Afrikaner speaks in justification of the government's activities, the matter of Communism in South Africa arises. When the Treason Trial commenced, Crown Counsellor Van Niekerk, prosecuting for the government, read excerpts from the Congress of Democrats

Bulletin, such as "freedom can be found only in Soviet Russia, China and the People's Democracies," and charged that the organizations of which the accused were members were dominated by Communists. Indeed, many of them had been members of the South African Communist Party before it was outlawed in 1951. The local Communist Party was founded in 1921, and at the last election in which it partook, in 1948, it got a total of less than 2,000 white votes. However, 2,000 active Communists could do much in South Africa through organizations like the Congress of Democrats, and each known Communist was kept under close surveillance. There is reason to believe that the Communists made considerable headway during the 1950's and early 1960's, which in view of conditions in South Africa is not unexpected. The African National Congress, particularly, seems to be infiltrated with Communists, although Luthuli is emphatically non-Communist. Luthuli, however, has never attempted to rid the Congress of Communists. There are estimates that between 1,000 and 1,500 active Communists are running various branches of the A.N.C., while others allege that between 60 and 80 per cent of all members of the Congress are Communist. The Natal Indian Congress is also said to be heavily Communist-riddled, while the Congress of Democrats long acted as coordinating body for resistance to the government. The movement of non-whites in South Africa is restricted, and thus the white Communists, able to travel freely, were valuable links between the various provincial organizations. Several members of the Communist party have actually been behind the Iron Curtain, and a secretary-general of the A.N.C., W. Sisulu, was in Moscow and Peiping in 1953. Congresses of the non-white organizations in South Africa have long been addressed by Radio Moscow and Radio Peiping, and it cannot be denied that there is contact between Asia's and South Africa's Communists.

It is, on the other hand, very easy to exaggerate the degree of Communist infiltration and the measure of success achieved. Notwithstanding the fact that South Africa's white army, air force, and police possess British and American equipment, that Apartheid is unfailingly hailed as the means of survival for Western civilization, that the history of conquest over the Africans is constantly recalled, and that the Communist nations attacked Apartheid more vehemently and much earlier than any Western country, there are still strongly anti-Communist Africans in South Africa. How fast their number is being reduced is not certain, but it would be a grave mistake to see all opposition to white rule in South Africa as Communist-led and Communist-inspired. It may be that the Communists are not failing, that they are merely content to await better opportunities to lead such opposition in the future. At present, however, there is no real evidence that Apartheid is a bulwark preventing the whole country from going into the Communist orbit.

Sharpeville, Black Week, and the State of Emergency of 1960 taught observers of South African affairs some important lessons. The Pan Africanist Congress and the African National Congress had not been able to agree on the methods to be employed in the resistance against the pass laws. After the shattering news of the killings at Sharpeville, there was a measure of unity among Africans throughout South Africa, and simultaneous mass protests occurred in all parts of the country. The police shot Africans in several towns, but it is a fact that tens of thousands of marching Africans, who seemed to be beyond checking, dispersed peacefully on most occasions when faced by a loudspeaker and police. The newfound unity soon began to decay, and there was fighting among Africans over the question of returning to work. The violence over, the police began to move against suspected leaders and sympathizers. People in Johannesburg, Durban, Pietermaritzburg, Cape Town, and Port Elizabeth were arrested before dawn in country-wide police raids. Lawyers, doctors, university lecturers, and others were held without bail or trial. A number were held incommunicado for some time. Others

REBIRTH OF THE REPUBLIC | 197

escaped over the Basutoland and Swaziland borders. Bishop Reeves of Johannesburg, who had insisted upon seeing the injured of the Sharpeville shooting, was deported. Non-white organizations were declared illegal. It was a crushing blow against resistance to the laws of the land, and in the exchange which started at Sharpeville and ended with these actions, the non-whites came off second best.

Having thus dealt with the first popular wave of agitation on the part of the non-whites, the Verwoerd government turned its attention to other matters. With the hint of deterioration in economic conditions, special efforts were made to attract investment. One such effort is the annual Rand Agricultural Exhibition, or, as it is popularly called in Johannesburg, the Rand Easter Show. The 1960 show was better than ever, and exhibits from countries all over the world drew hundreds of thousands of visitors. Prime Minister and Mrs. Verwoerd were on hand on the Saturday of the show's first week. As Verwoerd sat on the guest platform in the stadium, where 25,000 people were assembled to hear his speech, a man walked up to the Premier, pointed an automatic at his head, and began firing. The gun had been held only nine inches from Verwoerd's head, which was penetrated by two bullets. Pandemonium broke loose, the Exhibition was closed, regular radio programs went off the air, and South Africa was involved in a new crisis.

Miraculously, Verwoerd survived, and within days he was well enough to address the country in a message requesting calm. It was, fortunately, a white man who had shot at him, and not over a political matter but over a private one, involving some unsavory details. With virtually complete recovery assured, the shooting began to take on entirely another significance. Afrikaners, who readily invoke divine guidance as the justification for their every move, now hailed Verwoerd's recovery as a clear indication of the Lord's will with reference to his task. Verwoerd was back at work within weeks and announced preparations for the referendum to determine whether a republic was wanted by the electorate.

Isolation

With this, the days of the Union were numbered. On May 31, 1960, South Africa celebrated fifty years of Union with speeches and parades, but over the English-speaking section of the population there hung a mood of despair. For years, English South Africans had opposed many Afrikaner policies, but as long as South Africa was in the Commonwealth and ties with Britain remained as they were, there had been no desperation. Chaos in the Congo, trouble in Nyasaland, and independence celebrations for new African states north of the equator had meant little to South Africa. Now, however, there was a real chance that a section of the white population might lose some of its privileges—not at the hands of Africans but at the hand of fellow whites! South Africans of British descent were suddenly shaken out of their complacency, and for the first time there was something like a real campaign on an issue which actually and fundamentally divided South Africa's whites. Needless to say, this was not the sort of thing that helped South Africans forget their historical differences, and the old bitterness was back again. English-speaking South Africans were plainly aware that a republic might mean the end of South Africa as a member of the Commonwealth, which is something that also worried businessmen. The battle for the votes in this referendum gained momentum as the country was covered with signs saying YES–JA and NO–NEE, but the outcome was never in doubt. The referendum took place on Thursday, October 6, 1960, and the republicans won by some 850,-000 votes to 776,000.

These figures were probably the best indication ever provided of the relative strength of the Afrikaner and non-Afrikaner elements in South Africa. Once again, the Africans, Asiatics, and Colored were not consulted although they were deeply involved. However, Verwoerd had let it be known that he did not favor withdrawal from the Commonwealth, and this reduced opposition somewhat. There was now no doubt left that, among the whites, the republicans were well in the majority. It was decided that the new republic should be inaugurated on May 31, 1961, after the Commonwealth Prime Ministers Conference and on the last birthday of the Union.

Unexpected developments occurred at the 1961 Commonwealth Prime Ministers Conference. South Africa, as a republic, was to have applied for re-admission, and there had been speculation that India, Ceylon, Nigeria, or Ghana might veto such admission. But among these countries there was a feeling that the expulsion of South Africa might do more harm to the non-white peoples there than good, and the possibility remains that South Africa might have gained admission even as a republic. However, Verwoerd at the conference faced a barrage of hostile criticism of such intensity that even Macmillan was powerless to induce him to remain there. South Africa never applied for re-admission into the Commonwealth. Verwoerd withdrew the country voluntarily and with regret.

The country's isolation had never been more complete. Before, when there were still ties with the Commonwealth, South Africa could count on an insured trade area. To embattled South Africa, influential Britain was always a useful ally, and in conflicts with other Commonwealth countries, Britain served as mediator. Britain had repeatedly succeeded in modifying attacks of other member nations upon South Africa. Even if there was complete segregation in South Africa, at least at the annual Commonwealth Prime Ministers Conference white and black sat together, even South African white and African black. After the 1960 Conference, there had been some talk of diplomatic exchanges between South Africa and Ghana. Now that South Africa had withdrawn, the non-whites of the country knew that isolation could only lead to more excesses. If there were any checks on the government while

South Africa was in the Commonwealth, they would be removed. If there was to be an increasingly effective international trade boycott, the non-whites would be hit first by layoffs. In the Province of Natal, there was renewed talk of secession, and the Johannesburg Stock Exchange took another plunge. For the first time perhaps in South African history, emigration figures during the early part of 1961 exceeded those of immigration. There was minor rioting when Verwoerd returned from London, but the Nationalist press hailed the event as among the greatest of all time. The spirit of the Lager prevailed as it never did during Union days.

Truly, the country had never been so completely alone. Condemned once again in the United Nations for her racial policies and supported there—a doubtful honor indeed—only by Portugal's lone vote, censured for her actions concerning South West Africa, and boycotted in the world of trade, the Union during its last months stood in isolation. The Afrikaners accepted this condition self-righteously. Everyone else in the country viewed the grim future with apprehension, and many decided to leave. The Africans announced that a three-day general strike would be held to dramatize their objections to the course of events, and it seemed that the Union would come to an end amid greater strife and division than that which accompanied its birth.

The South African Republic, so long the dream of Afrikaners, became reality on May 31, 1961. On that day, Afrikanerdom achieved what it could probably not have achieved without the loss of the Boer War and without reluctant participation in the Union: a republic which extended from the Cape to the Limpopo. Not only were they boss in their old republican lands of the Transvaal and Orange Free State, but Afrikaners also ruled the Cape and Natal, where they were able to carry out their policies at will. Opposition had been crushed, the non-white masses remained subdued, and the Queen's Governor-General was relieved of his duties. The Afrikaners could savor complete independence, and could talk again of their President in the

Kruger tradition and could boast of the final defeat of their oppressors. Actually, in the new governmental organization, the position of President (given to C. R. Swart) is by no means as important as was that of Kruger in the republics of old. The power in the Republic of South Africa is held by the Prime Minister and his Cabinet and party. Thus Verwoerd remained, in effect, the country's leader.

May 31, 1961, was declared a national holiday. This date came on a Wednesday, and long before the end of May, African and other non-white leaders had stated their intention of striking from Monday, May 29, until Thursday morning. Mindful of the brief flurry of determination displayed by the Africans after Sharpeville, police began arresting leaders and threatening to shoot in the event of a general strike. In South Africa, non-white trade unions are not recognized by the government, and therefore strikes are not legal. Although the Nationalists confidently predicted that the strike would not materialize, there was a rush on stores at the end of the week preceding that of May 29. But the confidence of government and police was well founded. Strikes got under way here and there, but the great majority of the Africans went to work on Monday, May 29. There was virtually no violence, and once again, African organization and perseverance had fallen short of their mark, and police threats and action had done their work. With the wide advance publicity the strike had received, together with recollections of African masses protesting Sharpeville in 1960, the failure of the Africans was news which vied for space among the headlines with stories about the actual ceremonies inaugurating the Republic. The collapse of the strike produced a great deal of optimism among whites, especially the Afrikaners. Here was evidence that they were really *baas* in their own house and that the lack of control over Africans in such places as Kenya and Southern Rhodesia was not a feature of the new Republic.

There was little change in South Africa after the birth

of the new Republic. The Union Jack ceased to fly as an official flag next to the South African flag, although it continued to be hoisted in defiance in some places, particularly Natal. One more important change had already taken place months earlier: the English system of currency was replaced by a decimal system, and the Pound was replaced by the Rand, now the South African unit of payment. In Parliament, pictures of the Queen were taken down. New stamps were issued. But in matters of consequence, there was no change, except perhaps an acceleration in the application of Nationalist plans for the country.

Such calm may have reassured the Afrikaners and discouraged their opponents, but it revealed nothing about the future. There was little evidence in 1961 that the new Republic of South Africa would have an existence less troubled than that of its 19th-century predecessor.

FURTHER READING

ALLIGHAN, G. *Curtain Up on South Africa.* Boardmann: London, 1960.

BATE, H. M. *South Africa Without Prejudice.* Laurie: London, 1956.

LORD HAILEY *An African Survey.* Revised ed. Oxford University Press, 1957.

HATCH, J. *Africa Today and Tomorrow.* Praeger: New York, 1961.

HUDDLESTON, T. *Naught for Your Comfort.* Doubleday: Garden City, N. Y., 1956.

A Divided Culture

Language

South Africa is officially a bilingual country. Afrikaans and English have equal rights, and everything official is done in both languages. Official announcements are printed in each language separately, the courts permit the use of either language, and school is taught in one or the other, with lessons given in both. Every signpost, every train station, every post office, and every public facility must carry markers in two languages. After an uphill battle for equal recognition, Afrikaans has now fully asserted itself, which, in view of the power and prevalence of English, is one feature of which Afrikaners are justly proud.

Since a multitude of other languages are of course spoken in the country, to call South Africa bilingual is to understate the case. In Southern Africa, experts have recognized nearly 300 languages and major dialects spoken by Africans, and several of these are spoken by more people than speak Afrikaans or English at home. In South Africa, the most widely spoken African languages are probably Zulu, Xosa, Tswana, South Sotho, North Sotho, Tsonga, and Chivenda. The Asiatics also speak their own languages, and those most commonly heard are Hindi, Tamil, Urdu, Telugu, and Gujerati. Among the Cape Colored people, the Malay language has practically disappeared, and Arabic is taught in only one or two primary schools. The Cape Colored speak either Afrikaans or English.

The lack of concern displayed by the whites toward the African languages of the country is exemplified by the fact that African languages are only today making their entry into school syllabi. To become a government administration official, it was always necessary to be bilingual, but apart from some special cases, African languages were not required. Bilingualism itself was always a bone of contention among the whites. The Afrikaners take great pride in their language and resent the fact that many English-speaking South Africans do not know Afrikaans well and show hardly the slightest interest in learning it. It is probably true that proportionately more Afrikaners, through the years, have been bilingual. The number of non-bilingual people in South Africa is being rapidly reduced in one sense by the schools, which require a knowledge of both official tongues, but the continuing division between the two white sectors of the population has a limiting effect on usage. There has never been a successful two-language newspaper, and semi-official radio programs are broadcast on one transmitter in English, on the other in Afrikaans, although the commercial station uses both interchangeably.

Afrikaans is the subject of many misconceptions. Often defined as simply a degenerated derivative of Dutch, it is actually a language of beauty, individualism, and simplicity. It is undeniably an offspring of the Dutch language and has many words in common with Dutch. Nevertheless, it derives from 17th- and 18th-century Dutch, and just as Dutch itself has since gone through a lengthy evolution, so has Afrikaans. Marked changes in morphology and syntax have taken place, and the vocabulary shows a number of words derived from Malay, Hottentot, Bushman, German, French, and English. The Malay language as spoken by the slaves contributed certain words and elements of sentence structure. The impact of French is immediately obvious from the nasal sounds in the language. The result is a very distinctive language, which is perhaps somewhat limited in its powers of expression, but which is young and vigorous and possesses a growing literature.

One limiting feature of the language is the absence of the preterit; that is, in Afrikaans there is no equivalent for the words "gave," "went," "froze," etc. To say "I went" is the same as to say "I have gone" or "I had gone." Between all these, there is no distinction. To those acquainted with Dutch and German, this is rather surprising and perhaps the least satisfactory aspect of Afrikaans. On the other hand, there are also simplifications which enhance the language greatly. Unlike German, Dutch and French, Afrikaans has only one definite article and no complicated conjugation.

Afrikaans has been a written language only since the mid 19th century. For many years, Afrikaners themselves clung to Dutch as the official and respectable language of their country and saw Afrikaans as a sort of kitchen-Dutch, not good enough for official usage. In the second half of the 19th century, a society which called itself the Society of True Afrikaners began to work for the acceptance of Afrikaans (rather than Dutch) in the old republics. Initially, their success was moderate, there being no literature to speak of by which the capacities of the language could be illustrated. By the end of the century, however, much had been achieved, and a growing literature existed. The Boer War provided material and impetus for thinking and writing in Afrikaans, and by the time union came about, Afrikaans claimed equal rights with English in the new country.

Since those days, Afrikaans has become recognized as the independent language it is, completely separate from Dutch, and much "purer." The Afrikaners' feelings concerning purity apply to languages as well as race. Unlike the Dutch, who seem to consider the inclusion of foreign words, unchanged, into the Dutch language an asset, the Afrikaner linguists have proved the creative adaptability of their language beyond doubt. In the face of the power of

English, many had doubts that Afrikaans could survive. Their nervousness was not justified. During the 20th century, Afrikaans has gone from strength to strength, and it has been more than an equal partner to English in South Africa. A growing number of poets, novelists, and playwrights have made use of Afrikaans, and their work has gained recognition not only in South Africa but also in the Netherlands and Flanders. Such poets as Eugene Marais (writing in the decade of the Boer War), Jan Celliers, Van Wyk Louw (whose *Raka* must rank as one of the masterpieces of all time), Eybers, and a number of others have displayed the power of this language.

The development of Afrikaans as a modern language resulted in its recognition by universities beyond the borders of the country. Afrikaans was or is taught in the Netherlands, Belgium, Britain, and Germany, and its success became the subject of study by many philologists. In South Africa, however, Afrikaans is not always treated as well by Afrikaners as it might be. One reason for the undeserved reputation for ugliness which it has acquired is its everyday corruption by the general public. What makes Afrikaans picturesque and interesting is the absence of untranslated foreign words, but Afrikaners are likely to indulge in the use of borrowed words in preference to the Afrikaans equivalent. This is damaging to the language, and when outsiders hear the corrupt version, they ask whether a language containing so many anglicisms is really worth learning. Actually, the purity of "official" Afrikaans and the carelessness with which many citizens use it reflect an aspect of the character of the Afrikaner people. Afrikaans on paper and Afrikaans as spoken by the average Afrikaner differ greatly; more, perhaps, than in the case of, say, German or Flemish. The same is true of some Afrikaner policies, which, on paper, can be made to look genuinely fair as an effort to promote racial harmony. If carried out by everyone in a scrupulous manner, they might even have some success in this direction. Again, it is the average citizen who destroys the image and who twists the idea in committing gross discrimination. Afrikaners are also self-styled leaders of Christianity in South Africa, God's chosen people to guide the black man in the path of righteousness. Churches are crowded on Sundays, but the image is again tarnished, this time by the sale of some of the foulest literature—in Afrikaans—that is put on the market anywhere.

There can be no doubt that some of the negative aspects of Afrikaans, such as its yet limited literature, vulnerability to anglicization, and problematic nature when used in technical fields, may be attributed to its immaturity. In reality, the language is one of the products of nationalism, having a political significance whenever it is used. The irrationality and immaturity displayed by this nationalism are, not unexpectedly, found also in the Afrikaans language. As the shift of power has carried the nationalists from reluctant participation in the Union to domination of the new Republic in only a half century, so has Afrikaans asserted itself to take the initiative from English in this country.

The harvest of literary works of English-speaking South Africa is far less imposing than that of Afrikaans. English writers in South Africa had less stimulus and drive, and the country's output in this field has therefore been dominated by Afrikaans poets and novelists. However, some significant works in English have come from the pens of South Africans. Alan Paton's *Cry, The Beloved Country* is famous the world over, and other writers like Stuart Cloete have written novels on the South African scene. Nadine Gordimer, another product of English-speaking South Africa, has written soundly about the problems of the country.

The number of non-white authors writing in African languages is very limited in South Africa, but their number is increasing. Difficulties stare such writers in the face. If they write in their own language, then the market is limited. If they write in English, they are not writing as they would in their own language, which may have great power of expression. There is also the danger of official disapproval of anti-Apartheid writing by African novelists. Not long ago, an African author illustrated just how misguided some Africans in South Africa are when he wrote a little work (in English) called *Blanket Boy's Moon*, involving an African who flees the Union into Moçambique— the latter being called the "promised land"! The output of African writers in South Africa does not yet appear sufficient to permit an assessment of trends and characteristics.

❘ *Education*

Education in the Republic reflects both the assets and the liabilities of the country. South Africa is often described as a land of contrasts, and it is probably nowhere more true than in this field. For the Nationalists, education has been a means of indoctrination of white schoolchildren, and it is also a method of control over the Africans. For opponents of the Nationalists, educational institutions have been havens of refuge where in the name of academic freedom some severe criticism could go unpunished. Many of South Africa's liberals are or have been associated with the "English" universities, where segregation has been opposed most bitterly. The "Afrikaans" universities are dens of nationalism, although some faculty members have actually urged moderation of extreme views on Apartheid.

The schools, like the universities, are divided into those for Afrikaans-speaking children and those for English-speaking children. This separation has been made virtually universal by the Nationalists, although some single-medium schools did exist prior to 1948. The philosophy underlying this move toward the separation of the country's Afrikaans schoolchildren from their English-speaking contemporaries is dominated by fear of dilution of Afrikaner character and beliefs. The life of an Afrikaans child is governed by parents, teachers, and church people, all of whom, consciously, or unconsciously, indoctrinate him or her with Afrikaner ideals. It is undesirable that such a child should come into contact with children who belong to different churches, who have parents with different views, and who do not speak the Afrikaans language. There are, of course, administrative justifications for the separation of English and Afrikaans schools, but the urgent need for unification of the white population might have justified some administrative inconvenience. The inescapable conclusion is that the Afrikaner fears what he calls excessive contact between students from both sectors of the white population, and this fear perpetuates the division among South Africa's whites, even though they themselves recognize it as suicidal.

One of the notable features of education in South Africa is the white high school. Although there are differences among the Provincial Education Departments, the general approach to high-school education is along British lines. This remains true even though Afrikaners emphasize their dislike for things English and have toiled to rid their country of many of them. A child going through government schools starts with two years of kindergarten, after which there are five years of primary school, referred to as "standards" one to five. High school begins at standard six. There are five years of high school, leading to standard ten and "matriculation." This involves the writing of a final examination which is given to all final-year students in all schools. The writing of such a uniform examination everywhere insures equality of achievement and serves at the same time as an entrance qualification for the country's universities. A number of three-hour papers are written in each field in which the student has taken courses. The procedure resembles that in Britain and has all its shortcomings. There is general agreement that the standard of education in the South African high school is dismally low.

The two main reasons for the inadequacies of the high schools in South Africa are the nature of the syllabus and the standard of the teaching. There is no sound basis for the belief that South African high-school students are less capable than their contemporaries elsewhere and that this has depressed standards. They are made to work less hard and thus are rarely challenged to the limit of their capacity. There is some justification in the assumption that climatic and other environmental factors affect standards negatively, but there is nothing about the climate of Cape Town or Johannesburg that prohibits mental activity. Neither are the school facilities excessively limited; it is true that the libraries of most schools are inadequate, but improvements over the past years have not been reflected by comparably rising standards. Classes are not too large, buildings are generally excellent, and much money is spent on expansion everywhere. The trouble lies with the nature of the courses taught and the quality and attitudes of the teaching personnel. Indirectly, the universities and teachers' colleges—especially the latter—are to blame.

A large number of students leave high school after having completed standard eight, involving a bare minimum of requirements. Education up to this level, signified by a Junior Certificate, gives the student some knowledge of Afrikaans and English, history, and "science" (a mixture of chemistry and physics), but he may have filled out his curriculum with such fields as woodwork, which he may even carry to the matriculation stage. Students have to remain in school until the age of 16, but the drop-out at that age, with or without the Junior Certificate (standard eight), is alarmingly large. For those who carry on to the final examination at the end of standard ten, there are two years of advanced work in compulsory Afrikaans and English plus a choice from additional languages (German, Latin, or one of the African languages now included) and from fields such as biology (zoology and botany being included here), "science," mathematics, bookkeeping, woodwork, history, domestic science, etc. In addition to English and Afrikaans, four fields must be chosen as a minimum for matriculation. But any four will do, and a student may leave high school without knowledge of any language other than Afrikaans and English, without any contact with world history, or biology, or geography. In fact, geography was recently dropped entirely as a field of study by the Transvaal Education Department, ostensibly because of the overlap with history and because its inclusion in an interdisciplinary course in "civics" would render it superfluous. It is not difficult to find the real reason for the disappearance of geography from the high-school curriculum. Geography broadens, gives an insight into places and people other than the Transvaal, covers a number of topics which might include those not compatible with the South African government's present policies. Similar reasons can be given for the ridiculous neglect of world history along with over-emphasis of South African history; the results are obvious. The South African child is born into isolation, and this isolation is intensified by church, state, and school. It is untrue to say that the students who eventually qualify in biology and mathematics have not worked hard in these fields. They have, but while they did so, they lost their chance to learn to think. South African high-school graduates are incredibly ignorant of international politics, philosophies of government, the arts and culture. Few have really thought about race relations, and they are actively encouraged not to think in any other way than the government dictates. If a youth in South Africa, especially an Afrikaner, ever feels curiosity about such matters, it has disappeared by the time he is through with high school, or, to be frank, when high school is through with him.

The Afrikaner students, to be sure, learn more avidly and work much harder than their English counterparts, and the matriculation results each year emphasize their greater drive. Since the English schools, though they suffer from similar restrictions in the curriculum, are not involved so directly with the politics of nationalism, one would have expected them to produce better scholars. Perhaps the lower averages recorded by the English students are the very result of less cramming and more thinking. Certainly the higher institutions of learning in South Africa have produced eminent English scholars of world repute. On the other hand, the English schools, like those in

Britain, are often and justly accused of excessive emphasis on football, cricket, and athletics of other kinds. The better matriculation results of the Afrikaner high schools could thus be the simple result of greater—if misguided—application to work.

One of the problems confronting South African education is the variable quality of the teachers in the schools. Although there are many incentives for the prospective teacher to continue his studies to an advanced stage, the number who enter the profession with a minimum of preparation is alarmingly great. The damage done by poorly qualified teachers is bad enough when they give instruction in a subject they at least know some rudiments of. How much worse it is when, because of a teacher shortage, they hold courses in subjects with which they are barely acquainted even by name. Other teachers who are eminently qualified in one field are given tasks which prevent them from working in it. Recently, for example, one of South Africa's best teachers in the German language was made inspector of music for the Education Department, a field in which he was totally unqualified. Severe damage is done when, in the last two years of high school, a student is taught biology by a person whose real qualification is gardening or history by someone who is an amateur engineer and has merely a general teacher's certificate. Such things are, of course, not universal, but there can be no doubt that many South African teachers who pass for competent in the high schools do not belong there. Students have no way of indicating their displeasure, even if they were critical in these matters. One of the astounding aspects of the white South African schools is the manner in which the students accept whatever treatment they receive, be it brutal discipline, poor teaching, inconsistent grading, or ruthless indoctrination. They have no avenue of protest, and even the P.T.A. organizations never touch upon such matters.

The allegation that disciplinary measures in certain South African high schools are—or were until very recently—merciless is not unfounded. In high schools in Europe and

America, the student is accorded a certain measure of responsibility and freedom, but there is little of this in South African schools. Obviously, accurate descriptions of the over-all situation would have to be based upon acquaintance with every school, but a sample taken in the 1950's in the Transvaal produces some astounding revelations.

The control of the teacher over the student is absolute, and he may enforce his jurisdiction by the administration of severe beatings. There has been some moderation in this, and on paper the rule reads that only the school president may hit students, but teachers go unpunished for taking the law into their own hands. Practically anything can bring on a beating. High-school students in certain instances were, for instance, not permitted to talk while walking from one classroom to another between periods. They were ordered to walk single-file, and anyone caught talking could be told to report for a beating. A low score on an examination could likewise be punished by a thrashing. Other misdemeanors, such as noisiness in class, tardiness, loss of school materials, impoliteness, and the like could have similar consequences. The beating itself might take place in front of the entire class. One wonders what the effect on the beaten student as well as on the boys and girls in his class who are watching must be when he is told to bend over a table edge or chair, and the teacher, face contorted and coat flying, strikes him several times with a cane. On other occasions, students were taken to a small room and drubbed one after the other. Such beatings are not mild, and the injuries may be permanently visible.

It was interesting to note that the teachers who made most use of this procedure were also generally thought of as the least competent staff members. One, in particular, had the habit of walking around his class, asking questions. These questions, if they went unanswered, would be followed by fist blows (for the boys) and slashes on the hand with a ruler (for the girls). One student was knocked unconscious while the class was in progress, but this did not affect things in the least. It was extremely difficult to believe that in the second half of the 20th century students could be found anywhere who would accept such treatment or, for that matter, parents who would send children to such a school. Afrikaners, when confronted with the arguments against corporal punishment, frequently state that they are out to build men, and this builds men. Being able to withstand beatings, they feel, is a part of manhood, avoids flabbiness, and maintains the kind of discipline that is desirable in the Afrikaans high schools.

As a result of this, the atmosphere in the high schools in question is always tense, and the school years can turn out to be the least happy years in the students' lives. Particularly those less endowed with academic ability receive rough treatment, and they emerge from the school which set out to build men as bitter people, mentally immature and aware mainly of the usefulness of power. The frequent inadequacy of the teachers has hampered their efforts in schoolwork, and this has in turn resulted in frequent punishment. It is a system which could have survived only among Afrikaners in this form, with the teachers getting the approval of the parents and the local minister. To obey is one of the important lessons in the indoctrination routine, and there is no doubt that the student does learn this in his school. The school day begins (and may end, depending upon the teacher) with the "Our Father." Between the hours of 8:30 and 2:30, docility, blind acceptance of everything offered, is the order. The kind of respect shown to the teachers is a peculiar sort of fearful admiration such as can be found among African farm laborers on white farms.

The unhappy students can count on severe but inconsistent discipline. One morning, prior to the opening period, a teacher found a number of students sitting in the classroom, cramming. Announcing that there was a rule prohibiting entry into classrooms before 8:30 A.M., he administered a severe beating to all those present. It was subsequently found out that the teacher had broken up with his girl friend the previous evening and came to school in the morning with a hangover. One of the parents unearthed these and other unsavory details but found the school president and staff unanimous on the question: if the beatings were unjustified on this occasion, it was just as well that these 16- to 18-year-olds were cut down to size. Such is the stuff of which the Afrikaner mentality is made, and such are the conditions under which part of the new generation is being molded.

In addition to the regular courses taken by the students, all males are automatically enrolled for the weekly "cadet" (army) training course. One day each week, the teachers come to school attired as sergeants, and the boys have special uniforms. An hour's rigorous marching follows, and the Defense Department rotates army equipment among the schools, so that students learn to handle guns and ammunition. They learn fast, for the "sergeants" have the advantage of being able to administer physical punishment to the cadets right on the field. Some of the students go to special training camps during their vacation, there obtaining further training and some stripes. They become platoon leaders. Cadet training is compulsory, and no credit in terms of army duty is given for these years of drilling.

A small percentage of the high-school graduates enter the country's nine white universities. Of these nine universities, four use the Afrikaans language as the teaching medium, four use English, and one, the University of South Africa at Pretoria, is bilingual. This last institution is a correspondence school, which serves students not only in South Africa but throughout the continent and holds simultaneous examinations in places as far apart as Cape Town and Nairobi. There is an English university in Natal which has two centers, one at Durban and one at Pietermaritzburg. Like Natal, the Orange Free State has only one university, the Afrikaans Orange Free State University at Bloemfontein. The Transvaal and the Cape Province both have three universities. The foremost English university in South Africa is the English University of the Witwatersrand in Johannesburg. In Pretoria is the Afrikaans University of Pretoria, and in Potchefstroom is the Afrikaans Potchefstroom University. In the Cape, the famous University of Cape Town is English, and the corresponding Afrikaans university is nearby Stellenbosch University. In the Eastern Cape, at Grahamstown, is the English Rhodes University. Thus the Cape has two English universities and one Afrikaans, and the Transvaal has two Afrikaans universities and one English.

By American standards, the South African universities are not large. None in 1961 had much over 6,000 students, and most fell far under this figure. The influence of three of them has been felt throughout subsaharan Africa, however, for until recently the universities of the Witwatersrand, Natal, and Cape Town were unsegregated, and there are probably still more African graduates from them than from institutions of higher learning in Ghana, Nigeria, Sudan, East Africa, and the Federation combined. This lead is now being lost because the old universities have been segregated and the new "tribal" universities are developing slowly.

South Africa's universities are for the most part organized along the lines of the British system of higher education, and they suffer from many of the drawbacks inherent in it. By virtue of their linguistic separation, they draw people from opposing sides of the white population and are hotbeds of political activity. The universities of Pretoria, Stellenbosch, Orange Free State, and Potchefstroom are the "nationalist" schools of the country. Of these, Stellenbosch has the best academic reputation. This university has produced many prominent South Africans; Verwoerd, for example, was a professor there. In terms of organization this university has gone an independent course. There is a considerable amount of autonomy for each department, and student numbers have increased rapidly. It is significant, for instance, to note that while there is a trend against the teaching of geography in some nationalist circles in the country, the University of

Stellenbosch has a vigorous Geography Department which publishes a bilingual and unbiased journal for geographic education. This Department is organized along American lines, with teaching assistants and specialized staff, and it displays a refreshing, rather un-Afrikaans open-mindedness and academic objectivity and integrity. Other departments at this university are showing similar tendencies, and there is some encouragement to be derived from this. By contrast, the University of Pretoria has a poor reputation from the academic standpoint. Even though some years ago a number of professors at this university issued a statement urging moderation on the government, the attitudes of the vast majority of faculty and students are not open to any doubt. This is the core of the nationalist stronghold. Perhaps the most noteworthy feature of the institution aside from chauvenism is its association with the experimental research station at Onderstepoort, where some magnificent research has been done on agricultural biological problems. A similar comment is applicable to Potchefstroom University, which is also associated with an excellent station of this kind. It is always difficult to rank universities, but it is certain that Pretoria, Potchefstroom, and Orange Free State Universities are not among the leading four schools in the country.

The most famous institution of higher learning in South Africa—perhaps in all Africa—is the University of the Witwatersrand in Johannesburg. An outgrowth of a School of Mines (not unnaturally in Johannesburg), this university now consists of a large medical school, one of the world's best dental schools, a great engineering division, and departments of most of the sciences and arts. There are special divisions, such as the Bantu Studies Department and the African Climatology Unit, which make very important contributions to the knowledge of Africa. The Geology Department is one of the country's best, and the Geography Department is indisputably the leading department of its kind on the continent of Africa. Recently, much expansion has taken place, and magnificent new buildings evidence the prosperity of the institution. The new School of Architecture is a model of its kind, and the Geology Museum is the country's best equipped. Academic standards here are as good as anywhere in Africa, and a stream of highly qualified scientists is being produced. Although the government subsidizes all universities, the Witwatersrand University has worked hard to reduce its dependence upon government money; it constantly appeals for grants, and some years ago held a fund-raising festival to support its expansion plans. It has been singularly successful, and progress is evident everywhere.

The staff and students of the University of the Witwatersrand have long played a prominent role in the political affairs of the country. Unlike Pretoria University, its traditional rival, which is almost entirely Afrikaans, the University of the Witwatersrand is somewhat cosmopolitan. There are many Afrikaners or nationalist sympathizers in the School of Engineering, and the student assemblies at which representation is discussed display a divergence of opinions reminiscent of all South Africa. Formerly, when there were still numbers of Asiatics and Africans and some Colored people at this university, this was even more true. For years before the universities of South Africa were segregated, a sizeable minority at the University of the Witwatersrand kept liberal opinion alive. Since academic segregation was favored by far fewer people than favored social segregation, the enforced segregation of South Africa's integrated universities amounts to a tragedy. When the South African government first announced its intention of segregating the "open" universities, the strongest protests came from the Witwatersrand University's staff and students. Petitions were signed and protest meetings held, but to no avail. Although there was some delay, the present policy is that any attempt by a non-white to register at a white university constitutes a punishable offense. This places the onus on the would-be student rather than the university. It is not the university's task to keep him out; it is expected of the student to know the law and not to attempt to enter. Non-white students who had already registered and were engaged in studies at the time this law took effect were permitted to remain at the university until the end of their studies.

While it was a nonsegregated university, the University of the Witwatersrand was a showcase of hope for South Africa. There was very little friction among students of different races, and since nevertheless a fair number of the students came from a segregationist background, the prevailing harmony led to the moderation where moderation was most needed—doubtlessly a thorn in the flesh of the Nationalists. It must be emphasized that the University of the Witwatersrand was in effect only academically integrated. At athletic meetings, university dances, and other events of this kind, non-whites were never represented. Only the grill, in addition to the classrooms, was integrated. This is one reason why the principle of academic nonsegregation could be so jealously defended by people who favored segregation in all else. At "Wits," as this university is affectionately called, academic nonsegregation worked well, even in a country of Apartheid.

It is impossible to discuss academic segregation in South Africa without reference to the government's plans for higher education for Africans, Asiatics, and Colored. The final segregation of the "open" universities (Wits, Natal, and Cape Town) was based on another of the famed Acts, this one paradoxically called the Extension of Universities Act. While professors and their students marched in protest, plans were being laid for a number of brand new universities exclusively for non-whites. There was already such a university in the Eastern Cape, the University of Fort Hare, but this was closed down because of disturbances and suspected subversion there. New universities were begun, and no time was wasted. No time could in fact be wasted; with the segregation of the "open" universities, avenues of higher learning for non-whites were closed. As in many other matters involving Apartheid, the speed of the operation was and is unparalleled. Ordinary schoolteachers were taken from government high schools and given the rank and salary equivalent to associate professor at the new African universities, still in the building stage. Classes were started with the paint hardly dry and the facilities negligible. Verwoerd announced that these new colleges would be "better" than Fort Hare University ever was. While the first buildings were erected in one location, foundations were being laid elsewhere for additional schools.

The result is the ultimate in academic isolation. These new colleges for Africans are being built in the tribal areas, far removed from the towns and cities. Large signboards say that permission to enter the grounds must first be obtained elsewhere. An attempt is being made to incorporate into the architecture some Bantu building styles, and the languages used will ultimately be the local African tongue. For instance, the University College at Turfloop some distance from Pietersburg, is for the Sotho-speaking Africans of the northern Transvaal, and the teaching medium will be the Sotho language. In the vicinity of Empangeni, in Natal, the Zulu University College is being established, and here Zulu will be the language used. Other colleges are planned, and they will in time (if there is time) be staffed with Africans. Meanwhile, this is unquestionably a setback for Africans desirous of a higher education. They will no longer have direct contact with the best teachers in the country, and they will be separated from the best libraries and facilities; it would be ridiculous to echo the assertion that the new colleges will immediately have everything the older universities are able to offer. In some ways, they will never have the privileges of the white universities. The educational program to be offered is clearly aimed at local needs, and there will be little opportunity for the African student to interest himself in theoretical fields of study not considered to be in his own interest. The isolation will probably be permanent, and everyone knows the effect on a university of being away from the hub of things: members of the faculty of Natal University at

Pietermaritzburg talk of their isolation and the advantages Durban has—and Pietermaritzburg is just 50 fast miles from the bustling city of Durban! If Pietermaritzburg is isolated, what of Turfloop and Empangeni?

Lest the South Africans succeed in pretending that they are preparing for a wave of African matriculants who want to go on to universities, some figures might be quoted. Less than 3,000 African students succeeded in graduating from high school in 1960. In 1958, less than 600 students were accepted in teacher-training institutions, and less than 300 entered a university. The University of South Africa, always patronized by many African students, had about 300 correspondence students in the same year. For a long time, the scant number of Africans attending universities in South Africa equaled the number of Africans attending universities throughout the whole rest of the continent south of the Sahara.

Obtaining a bachelor's degree at South Africa's white universities is not difficult. It is, in fact, as easy as it is in Britain, the system being the same, and it is a source for constant amazement that the British should have managed to develop for their universities the reputation of rigor that they have. Americans who fear that Africans who have graduated from Wits, Makerere in Uganda, or Ibadan in Nigeria have nothing to learn at the undergraduate level in the United States may rest assured. These British-developed universities in Africa may be equal to London or Oxford, but they do not approach better American institutions. Undergraduates at reputable American universities work harder and longer than do their contemporaries in Africa (and Britain). There may be nothing at the American university to induce the meditation of the English student at Cambridge or Accra, but the absence of three-day cricket matches has always been seen as a great advantage. American students work four years toward their bachelor's degree, while South Africans (and Britons) spend three. An analysis of the South African academic "year" reveals the astounding fact that it lasts exactly 21 weeks, less than the length of two quarters of an American school year. Furthermore, the uncritical attitudes of the high-school students are carried into the university, and there is very little student criticism of courses and teaching methods. At one university a professor was reading the same notes he had used for over thirty years; at another a lecturer dictated at all times. Since there are no courses beyond the bachelor's level, the student merely writes a thesis to obtain an advanced degree. These theses reflect the inadequacy of the undergraduate preparation in all too many fields, and many would never be accepted for a higher degree at a good American university. With the sudden elevation of high-school teachers to university level, one shudders to think what quality of work will be done at the new African colleges or at the new Colored College in the Western Cape.

For years, South Africa's universities have been the professional homes of South Africa's most esteemed men. The "reaction at Wits," "comments from Natal," and "feeling at Pretoria" have long been accurate gauges of major lines of thinking on the country's problems. Many university people were and are politically active, but now they must be more guarded. The relative immunity from interference, so much a traditional privilege of the university professor, has ended in South Africa, and among those arrested when the Sharpeville shooting occurred in 1960 were prominent university lecturers. While all opinions could still be expressed at the universities, they were important outlets for all sectors of the population. Very many people paid close attention to what Professor Keppel-Jones, Professor Rautenbach, and others on opposite sides had to say. But the free voices are being muted, and as a result a disastrous exodus of university personnel has begun. As positions fall vacant, advertisements all over the world go unanswered. No country can afford to lose historians like Keppel-Jones, philosophers like Harris, geographers like Fair, linguists like Durrant, and doctors, dentists, engineers, psychologists, and innumerable oth-

ers. The emigration rate of qualified people at university
level has risen so sharply that some courses have been can-
celed altogether. The danger of a shortage of physicians
is drawing letters to the press from alarmed citizens. The
large majority of emigrants are English-speaking people,
but many Afrikaners are also leaving, regretfully, but for
exactly the same reasons. The future of the country is not a
pleasant prospect for academicians, whether they be of one
persuasion or the other, and younger Afrikaners figure
prominently among the departees. All this is happening
while the government constantly states that education is
one tool by which the white man can keep his lead over the
non-white. The stifling of academic freedom is having consequences in South Africa far beyond what was expected by
Nationalist as well as non-Nationalist observers. The withdrawal from the Commonwealth has caused the rate of
emigration to increase still further, and it will no doubt
have an adverse effect on the number of applications for
the newly vacant posts. Interference with education is stabbing at the country's heart.

| Religion

Afrikaners and non-Afrikaners alike are very religious
people. Among the Afrikaners especially, church and people are closely allied. The strength of the church helped
bind the Afrikaners of pre-Union days. It has acted as a
major guardian of the Afrikaans language. The dependence of the Afrikaners on their church has always been as
strong as it is today, and the church serves as moral justification for the Afrikaner's policies, aids in the process of
indoctrinating South Africa's youth, and is used as a political sounding board. Church attendance in South Africa is
high, and very few people are not affiliated. Non-affiliation,
in fact, leads to social isolation.

The latest available figures show the dominance of the
Dutch Reformed Church and associated churches in South
Africa. Of over 2½ million whites who attended church,

nearly 55 per cent belonged to these traditionally Afrikaans institutions. The next most important church, in
terms of attendance, is the Anglican Church, attended
mainly by English-speaking people in South Africa. It
draws only 16 per cent of the total church-going population. In descending order of importance are the Methodist,
Catholic, Jewish, and Presbyterian Churches. It is significant that out of this total of over 2½ million, only just
over 14,000 people indicated that they adhered to no religion. In addition to the major churches, a large number
of other faiths are represented. The Christian Scientists,
Baptists, Lutherans, Apostolics, and Plymouthists, as well
as Seventh-Day Adventists, Congregationalists, and many
more, are represented. There is religious freedom, although with the majority churches Catholics and Jews are
not popular. Among the Colored, the Dutch Reformed
Church and its associated churches are again dominant, but
less so than among the whites. Many Colored people attend the Anglican Church, and there are many more
Colored Congregationalists than there are white. The
Methodist Church is the other strong church among the
Colored people. The Asiatics mainly practice Hinduism,
to which about 70 per cent of the active churchgoers adhere, and Islam is second with just over 20 per cent. There
are some Buddhists and Confucians among the Asiatic
people in South Africa, but others attend the Anglican,
Catholic, or Baptist Churches.

Almost all denominations have been active in missionary
work of one kind or another among the Africans. About
40 per cent of all Africans do not today belong to any
Christian Church. Of the remainder, most (about 1.7 million people) belong to various Bantu Separatist Churches.
The Methodist Church has well over a million adherents,
and the Anglican Church well over a half-million. Although
the Dutch Reformed Churches have been active in the field
of missionary work (particularly of late), there are still
many more Africans in the Catholic and Lutheran
Churches.

The churches in South Africa, naturally, are deeply involved in many non-religious facets of life. The Afrikaners, feeling themselves the crusaders for Christianity among the savages of the dark continent, seek justification for their actions in the Scriptures, and when such justification is not readily forthcoming, they turn to their ministers for interpretative analyses of useful verses. Officially in South Africa there is separation of church and state, but in fact church and state are closely allied. The Dutch Reformed Churches are recognized as the government's churches, and once again the Afrikaners are unified while the non-Afrikaans whites are divided. The Calvinism of the Afrikaner has not mellowed much over the years, and there has been general—if largely unofficial—endorsement of the present policies of Apartheid. Particularly the government's activities in the field of education have been approved, and the church, in stating its support, has made it clear that there must not be "any mixture of languages, any mixture of cultures, any mixture of religions, or any mixture of races." Having "Christianized" many Africans, the Dutch Reformed church refuses to admit the new Christians into its houses of worship. Churches, like movie theaters and buses, are strictly segregated, and though the Anglican Church—among other churches—has attempted to break this form of Apartheid, it has had little success.

The Dutch Reformed Church, rather than keeping silent on the issue of racialism, has announced a policy where it might have been expected to avoid declarations of any kind. It refuses to associate itself with "the general cry for equality and unity in the world today," which it refers to as a "surrogate unity and brotherhood . . . in a world disrupted by sin." Churches should remain separate, it is felt, for such "independent churches can develop more fully . . . and reveal more completely the riches in Christ." Churches, pious Afrikaners intone, should not be established for the prime purpose of keeping devout Christians of different races separate, but this happens to be the result of independently developing churches. "Because of the

danger of being swallowed up by a numerically stronger
heathenism, which might have caused European civilisation
to lose its spiritual and cultural heritage, the Dutch Reformed Churches in South Africa did not hesitate to warn
against the integration of European and non-white races."
Such are the official statements made by the Dutch Reformed Churches' Commission for Race Relations. Although there have been calls for moderation by members
of the Dutch Reformed Church, and a split on the issue
has been developing for some years, this split is not sufficiently severe to disrupt the church. Whatever Afrikaners
say, they are Nationalists first and Christians second.
Where these philosophies can be made to correspond, harmony prevails. When it comes to a choice, nationalism
takes priority.

As in education, one can speak of Afrikaans-speaking
and the English-speaking churches in South Africa, and attitudes on religion and race relations differ considerably between the two. Although the English-speaking churches are
divided, they take a similar position with reference to race
relations and Apartheid and "affirm that the church
planted by God in this country is multiracial and must remain so. . . . This is one of its glories." This position is
particularly understandable when the non-white membership of these churches is considered; the Methodists, for
instance, have more Bantu ordained clergy than white. Although in their major assemblies these churches are interracial, ordinary Sunday services are carried out in a segregated manner. This is in part the result of the social
situation in the country; it is not possible to worship together each week and then to refrain from any other contact as which the law prescribes. The Anglican Church, and
other churches, protested strongly against the removal of
the Colored from the Cape common voters' roll and
against education policies of the government. But the doc-
trinal statements concerning integration and its desirability
have not been accepted by a large section of the white membership, and little has been achieved in terms of moderation

of prevailing conditions by these English-speaking churches.

In their fight against discrimination, several church leaders have been thrown into national prominence. Father Trevor Huddleston, worked in Johannesburg for years for the advancement of non-whites, became involved in a series of conflicts with the authorities and wrote up his experiences and opinions in *Naught for your Comfort* after leaving South Africa permanently. Admiration for Huddleston is mixed with disappointment at his periodic inability to judge matters with a degree of objectivity; he opposed some measures which actually benefited the African, just because they were Nationalist-initiated. Immediately after the Sharpeville shooting, Bishop Reeves of Johannesburg insisted upon seeing the injured, reported on their condition, and, when the wave of arrests began, fled to Swaziland. Upon his return, he was deported. Long an outspoken critic of the Nationalists, Reeves is one of the most able churchmen ever to work in South Africa, and his expulsion removed a strong voice from the national scene. An Afrikaner minister, G. J. J. Boshoff, who formerly had a congregation in suburban Johannesburg, has done much to associate the church with politics. His fiery sermons would do justice to any political arena, though they often have little to do with religion. An extreme Nationalist, he was expected to exchange the pulpit for the political platform, but he remained in the church. Boshoff has not enhanced the image of his church, but the contrasts between his attitudes and those of Reeves and Huddleston indicate just how far apart the Afrikaans and English churchmen in South Africa are.

Missionary activities and fragmentation have led to the large number of African Christian Churches now in existence in South Africa. Their very number creates problems, and repeated calls have been made for the unification of Bantu Christians in one comprehensive Bantu Christian Church. Such an appeal was, for instance, issued by the President of the African National Congress. It was rejected by African church leaders, and this rejection has become a

major justification for the continued segregation of
"white" churches from the Bantu churches. If, the whites
argue, the Africans themselves do not wish unity in religious affairs, what reason is there to promote interracial
unity in the church? It should be noted that the call for
religious unity came from a politician and was rejected by
churchmen. Religious division is one of the features of the
general disunity among Africans and is an obstacle to
politicians who seek African solidarity. To the many
African churchmen, consolidation of all churches brings
with it the danger of loss of stature which some have attained in the institution and the possibility of limitations
placed upon the manner of worship. Religion is the only
area in which Africans could establish an organization
without fear of outside interference, and it is obvious that
those who have managed to build a small empire are not
willing to let it be amalgamated. As a result of this freedom and in addition to the major churches, about 1300
Bantu sects exist, many of which can be described as semi-
Christian. These sects are not accepted by the Christian
Council of South Africa or by the World Council. Many
are drifting into paganism and witchcraft, and from the
point of view of Christians, the education of the leaders
of these sects is a matter of urgency.

Ever since the days of the London Missionary Society—
and in a less organized fashion, even earlier—white men
have been attempting to convert Africans to Christianity in
Southern Africa. Portuguese, Swiss, Dutch, British, German, and French missionaries, as well as a few Americans,
have toiled to spread Christianity in Southern Africa.
Today, after a lengthy battle against those within the
church who felt that such dispersion of Christianity was a
danger to white civilization, the federated Dutch Reformed Churches in South Africa strongly encourage mission work with substantial grants. As a result of all this effort, a majority of Africans in South Africa now adhere to
one or another of the Christian churches. Whether Christianity suits the African is a matter for debate. In many

places, polygamy is an essential part of the social and economic system. Renouncing three of four wives may not be only a matter of social inconvenience for an African; it may be economic disaster. From the northern Transvaal and elsewhere come sad tales of the effects of conversion. To some observers, it will always remain a surprise that Christianity has had the impact that it can today claim. Yet it is with this least tangible, least certain of the white man's possessions that practically all Africans have had some kind of contact. There is little in South Africa today that can convince the African that the white man lives the Christianity he preaches—and yet millions of Africans are devout if unhappy Christian people. The excitement and color of pagan ceremonies have been replaced by the dismal whitewashed church and off-tune choir. Monogamy has been imposed where for centuries polygamy was the practice. Discrimination exists on the basis of the idea that the unity of the Christian church does not mean the equality of its members, and yet Africans embrace this white man's gift.

Whether Christianity will survive in Africa long after the termination of white rule everywhere is another matter for contemplation. Islam is making inroads in Africa north of the equator, and its attraction lies largely in its non-discriminatory racial practices and its non-association with the white man. In the years to come, these ought to be powerful weapons in the competition with Christianity.

| The Press

The South African press enjoys a surprising amount of freedom. Criticism of the government, ineffective as it may be, can be voiced by any newspaper. Newspapers have indeed been banned, especially African-produced publications, but the anti-government papers such as the Johannesburg *Star, Cape Argus,* and *Natal Mercury,* can count on a high degree of editorial immunity. South African journalism is of good quality, and the majority of the

newspapers are a credit to the country. Generally, the press
uses its freedom with some responsibility, which no doubt is
one reason for its survival. There has been periodic interference with news reporting of such events as the Sharpeville shooting, but there is no wholesale censorship at any
time. Americans and Britons would perhaps not consider
the South African press as free as it might be, but Portuguese, Spaniards, and citizens of many other countries have
never read papers which object as strenuously to government policies as do those of South Africa. In this respect, South Africa is head and shoulders above some of
the "free" nations of the world.

There is no national newspaper in South Africa. Communications are not sufficiently fast in this vast country, and
so each major city and town has its own competing group
of dailies. Practically without exception, in the metropolitan areas there are an Afrikaans and an English morning paper as well as two evening papers. The Argus Print-
ing Company, dominated by mining interests, publishes
three major English-language dailies in the cities: the
Star in Johannesburg, the paper which comes closest to
having a nationwide readership; the *Cape Argus* in Cape
Town; and the *Daily News* in Durban. Each has its Afrikaans competitor. Other important English-language papers are the *Rand Daily Mail* in Johannesburg, published
in the morning; the *Cape Times* in Cape Town, which has
strong competition from the only really adequate Afrikaans paper, *Die Burger;* and in Durban, the *Natal Mercury*, which was once known for strong opposition to
Smuts. In Pietermaritzburg, a minor paper which publishes
some of the best editorial material to appear in the country
is the *Natal Witness*. The Afrikaans papers are a disappointment, with the possible exception of *Die Burger*. Second to the Cape's *Burger* is the Johannesburg morning paper, *Die Transvaler,* virtually a government paper which
has always followed the Nationalist line without deviation.
The Afrikaans evening paper in Johannesburg is *Die
Vaderland,* and it is one reason for the *Star's* large circu-

lation. As might be expected, there is no major Afrikaans daily in Natal and no significant English paper in the Free State. Only in Johannesburg and Cape Town is there any real competition for the market. Smaller places, like Port Elizabeth, East London, and Pretoria, have their own daily papers, and dailies thrive on amazingly small circulations. Circulation is to be measured in the tens of thousands only, for South Africa's press is a regional press.

In addition to these prominent papers, there are a number of minor publications, newsletters, and the like, all of which have to be registered at the Post Office. This is why South Africa boasts the amazingly large total of some 600 newspapers. Some of these are the Sunday papers, such as Johannesburg's *Sunday Times* (published by the *Daily Mail* and not among the high-quality papers of the country) and Durban's *Sunday Tribune.* Others are published by Africans, among these the *Golden City Post* and the *New Age,* the latter being accused of following a Communist line.

Although some of the publications among the 600 issued in South Africa are of poor quality, part of what is still called locally the Yellow Press, South Africa's major daily papers never sink to the level of the British and American tabloid press. There may be no *New York Times* or *Christian Science Monitor* in South Africa, but the *Star* and several of its contemporaries are among the better products of journalism. Editorially, matters of importance are dealt with in a competent manner on both sides. In terms of reporting, the newspapers all have access to the reports of S.A.P.A., the South African Press Association, and world news is slighted no more than it is by many prominent newspapers in the United States. Naturally, both Afrikaans and English papers show a bias in the interpretation of events in South Africa. Afrikaners feel that the English press displays irresponsibility in its destructive criticism of South Africa and that some reports, which are read overseas, are untrue and damaging to the country's reputation. Actually, some blunders have been committed by English papers. In their efforts to discredit the Nationalists, English-language journalists have come up with sensational exposés about ill-treatment of Africans, only to see the stories proved false when put to the test by the Afrikaans papers. English-speaking South Africans criticize the Afrikaans press for its utter lack of independence and fairness. The Afrikaans press is produced by and for Afrikaners. It stands for Afrikaner nationalism, church, and language and knows little moderation. An average issue contains mostly political news, profiles of leaders of the Party, attacks on the Opposition, and local material; foreign news is much neglected. Reports of attacks by Africans on whites are played up, as are such disturbances as the liquor riots in Durban. Church news has a prominent place. The daily papers, unfortunately, do not give the Afrikaans language the treatment it deserves and thus play a role in its deterioration.

More Afrikaans people read English newspapers (often in addition to the Afrikaans daily) than English-speaking people read Afrikaner opinion. This was probably more true in the past than it is today, with bilingualism increasing rapidly. For many years, however, when Afrikaners were able to read the English press, the English-speaking public was virtually cut off from Afrikaner opinion by the language barrier. Even English newspaper editors were for years unable to read the products of their Afrikaans competitors and could not understand Afrikaans government ministers at conferences. This is changing, although it has not been eliminated. In their bitterness over events in recent years, many English-speaking people refuse to learn Afrikaans or, if they have been compelled to learn it in school, decline to use it later. There are still people on the staffs of newspapers in South Africa who are not fully bilingual. The reasons for the greater circulation among Afrikaners of the English papers are their completeness in terms of news-reporting and the fact that they are, really, quite moderate. History has something to do with this. The Afrikaans press takes a Nationalist, extreme, dogmatic attitude. Its major objective of long-standing lies in attacking the Opposition particularly, labor, liberalism, communism, integration, and a host of other things generally. The English press, especially the Argus papers, which have in a way set the pattern, can be neither too liberal nor too mild. It cannot be excessively liberal, for in its history as an organ of the Chamber of Mines, the Argus Company has had to avoid the wrath of the government. Who wanted to increase taxation on gold production? Neither can it afford to seem pro-Afrikaans, for it has a readership ranging from white to Colored and Asiatic; more and more Africans can also be seen to carry the *Star* home. In appealing to the non-whites, these papers have the great advantage of language.

Genuine efforts are made by both the English and Afrikaans dailies to present opinions from the other camp. Editorials are translated and printed, and commentaries are exchanged. All this helps to prove just how little division on essential matters will be found ultimately among whites in South Africa. There is—bar certain non-white publications—no liberal publication in the country. Neither, surprisingly, are there any columnists of national reputation. There is no James Reston in South Africa, no journalism remotely approaching the nature and quality of his work. There have been one or two journalists in his tradition in the past, but there is none at present. At independent papers like the *Natal Witness*, some individuals with liberal convictions are active, but they do not have an impact on the national scene.

The African press has had a difficult time recently, but publications banned by the government keep appearing under another name. Some are quite vehement in their criticism of events in South Africa, and foreign readers will be surprised at what can publicly be said in the country. When Viscount Montgomery arrived from Britain some years ago and started making pro-government statements, he received some rough treatment at the hands of the non-white press. When, recently, there was trouble in Pondoland in the southeast of the country, and police, army, and navy units were active in combating it, the African press demanded information on the whereabouts of those arrested. In comments concerning Sharpeville and Black Week, calls for African action could be read, and papers which printed statements of this kind are still publishing. The freedom of the African press, however, is far from complete. Editors and journalists have been arrested and jailed, and reporters are frequently harassed by police and detectives. Some papers have been suspended or permanently banned. Nevertheless, in a country of so much control over so many aspects of daily life, the degree of freedom of the press remains a pleasant surprise. Perhaps some would prefer the interpretation that in spite of all the restrictions the government attempts to place on the press, particularly the non-white Press, there has been some success in expressing most shades of opinion, even those beyond liberalism. The result remains an unexpectedly varied set of newspapers.

| Broadcasting

The South African Broadcasting Corporation (S.A.B.C.) holds the monopoly on radio in South Africa. It operates a non-commercial system in English as well as one in Afrikaans, and in addition the Corporation has a commercial station called Springbok Radio, on which Afrikaans and English are used interchangeably. The noncommercial programs are set up along the lines of the British Broadcasting Corporation. Live plays, symphony orchestra performances, lectures, debates, and variety programs are transmitted, interlarded with several complete newscasts during each day. The broadcast schedule does not call for a long day. On the regular programs, sign-off comes just after 11 P.M., and the commercial station goes off the air at midnight.

In addition to its national transmitters, the S.A.B.C. has

an intricate organization of regional stations. These are used to give local news after the bulletin of national and international news. Sunday church services and evening recitals are also broadcast in this manner, so that listeners in Natal can hear a Durban or Pietermaritzburg program, those in the Cape one from Cape Town, and so on. In addition, on certain hours of the day, the regional transmitters use African and Indian languages in special programs aimed at the African and Asiatic audience. S.A.B.C. broadcasts in Zulu, Hindi, Urdu, Tamil, Telugu, and Gujerati from Africans and Asiatics in Natal, in Xosa for the Eastern Cape, in North Sotho for the northern Transvaal, and in a number of other African languages for other regions of the country. In many African suburbs of Johannesburg, there is a radio distribution network, a service which places a loudspeaker in the home (but no set) and which is tuned constantly to the S.A.B.C. Otherwise, the S.A.B.C. is received by normal radio sets on several wavelengths. The station at Lourenço Marques (a commercial station in neighboring Moçambique with strong transmitters which reach all of Southern Africa), Cairo Radio, Radio Brazzaville, and other stations can also be heard. Because of interference from these outside stations, the S.A.B.C. is now developing a 35-million-dollar network of frequency-modulated stations.

The people who broadcast to Africans are themselves Africans, and the S.A.B.C. is a multiracial venture, although it is in white hands. The Africans who cooperate in the Corporation are "loyal," and there are a number of African newsmen in the service of radio, as well as over a thousand white correspondents all over the subcontinent. During Black Week in 1960, some efforts were made to induce African radio personalities to appeal for agitation, and there were one or two suspensions. But in general, the S.A.B.C. does not editorialize much on politics and concentrates on producing "good" programs. In this, it is felt, the Afrikaans section of the organization is more successful than its English counterpart. Afrikaans programming suffers from only one really unpleasant feature, and that is the interminable hours of "boeremusiek," a sort of hillbilly music seen as a national art form by many Afrikaners. The music itself is not so unpleasant as are the renditions of it by the bands called upon to perform. Everyone in South Africa seems to be on the boeremusiek bandwagon, and any number of groups play it—most atrociously. As a result, people who should spend a few years practicing and a few more confining themselves to amateur talent competitions are put in front of an S.A.U.K. (Afrikaans for S.A.B.C.) microphone. On the other hand, the S.A.U.K. far excels the S.A.B.C. in the promotion of serious music, chamber music, recitals, and lectures. Its programming includes much less light material than does that of the S.A.B.C., and the effort to elevate standards is everywhere evident.

Improvement of standards is not something in which the S.A.B.C. in general has had a great deal of success. It seemed that a sizeable step forward was taken when the S.A.B.C. Symphony Orchestra was merged with the Johannesburg City Orchestra, but preposterous hiring policies caused the most able instrumentalists to depart for Europe or to take up positions elsewhere in Southern Africa. Nonetheless, the orchestra for some time constituted an improvement if only in terms of size—it was the only full-scale symphony orchestra in the country. Very soon, however, the lengthy working hours and pressures of frequent performances began to tell, and deterioration set in. Today, the orchestra is struggling to maintain membership in all sections, and is failing. Without unusually good salaries and working conditions, South Africa is no more attractive to musicians than it is to university professors. The orchestra remains one of the Corporation's most extravagant luxuries, but the mediocrity of its membership requires excellent leadership for adequate performances. Since, with the exception of the guest-conductor season, local and inexperienced conductors are always at work with the orchestra, strong, competent leadership has always been absent. As long as this remains so, this orchestra will not perform satisfactorily.

The commercial broadcasting subsidiary, Springbok Radio, has been in operation just over a decade and has proved highly successful. It is comparable to American A.M. commercial stations, advertising soap and refrigerators, aspirin and tea. Recently, news commentaries were included with the scheduled programs, and this gave rise to some friction during Black Week, when Springbok Radio's newscasts contradicted those of the S.A.B.C., whose every word was carefully planned so as not to arouse excitement. Springbok pulls its news from the wires of the Associated Press, while the S.A.B.C. relies largely on the South African Press Association. Springbok has been ahead of the S.A.B.C. on several occasions, and there has actually been some internal competition, with the healthy result that the Corporation's news section is receiving added attention.

The South African Broadcasting Corporation performs a peculiar function in the Republic. While not officially a government organ, its newscasts and commentaries are designed in such a manner that no unfavorable light is shed on local affairs. In recent elections, radio was for the first time employed by political leaders to state their case. Equal time was granted to the parties, and sedate, useful speeches were read by the major figures in the political arena. Other than this, however, the S.A.B.C. plays no active role in the political affairs of the country, and it cannot with justification be called the mouthpiece of the government. On the other hand, what is said over the programs for non-whites is zealously checked and must be approved. The government subsidizes the S.A.B.C. and hence has, at any time it becomes necessary, vital control over its functions. Meanwhile, the Corporation enjoys a measure of independence and attempts generally to present programs of a non-political, non-controversial nature.

South Africa is one of the few economically advanced countries in the world which does not enjoy the "benefits" of television. Steps are being taken in the establishment of television programming, but for years these efforts have been vigorously opposed in government circles. There are fears that the country is unable to afford television, but in localities such as the Witwatersrand where several million people could be reached, this is hardly believable. There have been questions concerning the effect of television on morals, and particularly the effect on African viewers of shows containing violence. One of the problems is how to control television once it has been established. In neighboring Moçambique and Southern Rhodesia, plans are in an advanced stage, and television—whether South African or not—will, like outside radio stations, penetrate at least the fringes of this country kept so meticulously isolated. In the field of mass communication, there is no free enterprise. Television, if it is introduced, will be government-controlled like radio. Probably all television sets, like all radio sets in the country, will have to be licenced with permits which must be renewed annually. Thus, no outside corporation has been free to make an attempt at the establishment of television in South Africa, and so far the development of this medium has been thwarted.

| The Arts

There is an active and growing film industry in South Africa, supported by a group of very enthusiastic individuals who have produced some noteworthy pictures. Several of South Africa's writers have sold their works to be filmed, and Alan Paton's *Cry, the Beloved Country* was among the first. One of the earliest films in Afrikaans was one based upon the famous writer D. F. Malherbe's *Hans die Skipper* (liberally translated as Hans the Fisherman), which was a significant work in spite of the adverse criticism it received, particularly in the English press. Since then, a number of cheap comedies have been distributed, but also some excellent short films and one or two major efforts which have evidenced the growing maturity of the industry. Other South African writers have sold their manuscripts to overseas film companies. Cloete's *Fiercest Heart,* for example, was bought abroad and filmed in Natal.

There are also a number of small but very ambitious repertory companies in Johannesburg and the other cities. They have struggled against public indifference and have achieved a measure of success, but even in Johannesburg today plays never run as long as they deserve and rarely receive the support they so badly need. This is one of the real paradoxes of the South African scene. Afrikaners who are so proud of their language and literature fail to support one of its most significant manifestations: live theater performances of Afrikaans authors' works. Yet some excellent plays emerge which deserve a better fate than a two-week run in the Library Theater. The professional National Theatre does them justice: some great yet unheralded actors work on the South African stage. The dean of the Afrikaans stage, Andre Huegenet, died in 1961, and his death marked the end of an era for Afrikaans theater. Huegenet was once a nationally acclaimed figure, but he went out of favor with the public and faded from the scene. Among the other notable stage personalities, Siegfried Meinhardt takes an important position.

There have also been repeated efforts to establish a permanent local opera, but these have not met with much success either. In the better years, local artists perform with the Johannesburg City Orchestra, but the annual visits of Italian opera companies get the public's preference, not unnaturally. It cannot be denied that the performances by local artists at times leave a good deal to be desired, but much effort went into the staging of the *Marriage of Figaro* in Afrikaans, *Carmen* in English, and so forth; it is regrettable that the press saw fit to criticise these productions so severely. Without local support, local talent gets nowhere, and if it is good talent, it goes overseas. South Africa has lost opera stars of international quality, instrumental soloists who are among the best anywhere, ballet dancers who have taken their places among the leaders of world famous companies. Naturally, the very best would have entered the world circuit whatever their home country. But the sound, merely good performers leave for lack of being able to make a living.

Some years ago, Menotti's *Consul* was produced by an all-South African cast and orchestra working under adverse conditions in an inadequate hall. The result was highly creditable and was in any event a great service to the South African public, which had the opportunity to hear a modern opera rarely performed anywhere. What was particularly encouraging was the composition of the cast: an Afrikaans soprano sang superlatively, other roles were taken by English-speaking South Africans, and the orchestra contained local players as well as musicians who had immigrated. The opera was received with a great deal of enthusiasm by the public on opening night, but the *Star* declared its quality to be poor, audiences dwindled, and another attempt to encourage South African arts had been dealt a severe blow.

Without a doubt, the *Star,* more particularly the man responsible for its cultural page, Oliver Walker, has done incomparable harm to South Africa's reputation in the field of the arts. When the Collegium Musicum Trio came to play in Johannesburg, it was accused of having an improper name; it was not, announced the *Star,* connected with any college (*sic!*). To those who have penetrated beyond the limits of the English language—as one might expect a critic of cultural events to do—*collegium* means, among other things, ensemble, a good name for a string trio. Writing of this kind has infuriated and inhibited South African performers, while at the same time persuading musicians from overseas not to visit the country. Other papers have just as bad an effect by blandly praising all cultural affairs of whatever kind or quality. Except for the *Rand Daily Mail,* which has long had a very able music critic, all too many papers simply describe *every* performance as excellent or good. Little or no attention is paid to these Pollyannas (reviews in the *Natal Daily News,* for example, are ignored), and the field is left open for

Walker's effusions, which go to the other extreme. This is another example of the power the press can have over local cultural affairs and of how important it is that such power be used responsibly and, above all, competently.

While performances in South Africa may at times leave something to be desired, there is ample justification for the pride the country takes in its composers. For many years, South Africa was virtually without artists writing instrumental and other serious works, but during the period from 1916 to 1927, four composers were born who have provided a group of works which form the basis for what may be a bright future. The oldest among these men is Arnold van Wyk, whose symphonies have attained a reputation extending beyond the borders of his country. Hubert du Plessis and Stefans Grove have written chamber music and piano and vocal pieces as well as orchestral works. The youngest member of this foursome is John Joubert, who has composed for piano and orchestra and has written cantatas and songs. These relatively young composers enjoy an international reputation, and their works are performed in orchestral programs in many parts of the world. As might be expected, they make abundant use of the local idiom, both indigenous and non-indigenous, and traditional tunes and rhythms of all kinds find their way into chamber music and symphonic works alike.

In the field of painting and sculpture, South Africa has produced noted artists. Probably the most famous of all is Pierneef, whose paintings of the South African landscape are among the most significant works to come out of this country. His experiments in cubism served to introduce an entirely new facet of modern art to the local public, and so characteristic are his interpretations of the landscape of the South Africa he knew so well that today people talk of a "Pierneef tree," "Pierneef sky," or "Pierneef mountain range." A prolific and tireless worker, he spent a lifetime producing for South Africa a group of works which have brought fame to the country in art circles. Many followed in his footsteps. Pierneef was a pioneer, born in Pretoria, who as early as the 1920's, when art in South Africa was in a stage of youth, painted works which have become recognized as among his most important. South Africa in those days was not willing to accept the modern and contemporary approach, and another pioneer, Kibel, suffered even more ridicule than did Pierneef and other artists, but without achieving eventual success.

In recent years, the number of artists and the diversity of their work has increased, and the flavor of Africa has become increasingly noticeable in what has been produced. The value of the heritage of Bushman paintings has been recognized, and sculptors and painters alike have adopted a new national trend, in part based on this and other indigenous traditions. Maggie Laubser, much of whose work was done overseas but who also joined in this new development, Preller, Battiss, and Shephard have turned out significant paintings, while sculptors like Lipschitz and Steynberg have displayed the African touch in their field of creative art. Naturally, the entire movement still displays signs of immaturity, but it is vastly interesting in its vigor and vitality, the sudden sureness of direction and individuality.

| Sports

White South Africans are avid sportsmen. With the leisure time they have available and the good climate of their country, their abilities in this field are not unexpected. The national winter sport is rugby, which is remotely comparable to American football, though much less intricate and requiring less skill. Competing with increasing success with rugby—in which the South Africans long held the unofficial world championship—is soccer. There is now an interprovincial professional soccer league, and while formerly many good players went to Italy and England to play for money, they now have the opportunity at home. In summer, cricket is the major spectator sport, and South African teams receive the national teams of Australia, England, and New

Zealand, while they themselves frequently tour overseas. Of course, play with the West Indies, Pakistan, or India is out of the question because of racial discrimination.

In South African sports, as in everything else, there is a rigid color bar. Very rarely, a white team plays a non-white team (in Natal this does still happen), but it is frowned upon, and there are no mixed teams of any kind. Thus there are African soccer leagues, Asiatic cricket leagues, and Colored tennis matches but no interracial competition. There are constant objections from the non-white sport associations, and at the last Olympic Games a protest was lodged concerning the composition of the (all-white) South African team, but there is not likely to be any change. Hence, international teams from South Africa represent only one small part of the population, and they could be even stronger if representation was put entirely on the basis of ability. It must be admitted that if such competition between white and non-white occurred, the whites would at first take almost all places on any team. One sport in which Africans have achieved prominence is boxing, but to achieve success a non-white boxer must travel overseas for his fights. He cannot defend his title at home against white challengers. It is always interesting to observe the interest and pride South African whites take in the success of a non-white who competes overseas, apparently forgetting his inability to rise to the top in his own country because of laws they themselves support.

In Johannesburg is the headquarters of soccer, the Rand Stadium, nestled between the mine dumps of the south. Here, upwards of 25,000 people will attend games between English and other foreign soccer teams and representatives from white South African soccer. In this stadium there is a small section reserved for non-whites. It is confined to the southeast corner of the stadium, the worst possible spot. But the few Africans and other non-whites who can squeeze into it virtually to a man support the visiting team, take every opportunity to harass players from the home side, and demonstrate their support for the visitors with flags, banners, and an unbelievable noisiness. It is something of an experience to see Africans from Zululand who work in the Transvaal gold mines waving the Union Jack around in support of a soccer team from Newcastle, England. More police are often required to keep this section of the stadium under control than the entire remainder, and there have been some ugly fights between occupants of the adjacent white sections of the segregated stadium and this non-white block.

White South Africa has produced world champions in boxing, hurdling, and swimming and has sent finalists to Wimbledon and other major tennis tournaments. Bobby Locke and subsequently Gary Player have kept South Africa among the leaders in golf. For a short time, baseball caught on in South Africa, and large crowds attended the games between the American All Stars and local teams, but the interest has decreased lately. The sport in which more people participate than any other, however, is bowling. This game, which is played on outdoor lawns, is not related to the American variety. The object is to roll an imbalanced, oval-shaped ball, called *wood,* as close as possible to a small white ball called *jack* which is placed at the end of the lawn or *green.* It is a game that can be played by all, young and old alike, and its popularity is immense. There are possibly more participants among white South Africans in bowling than in all other sports combined. Formerly seen as a game mainly for older people, it has begun to appeal to younger players also, swelling numbers even further. Among Africans, however, the sport is virtually unknown.

Horse racing is a popular pastime in the Republic, and there are large tracks in Johannesburg, Cape Town, Durban, Pietermaritzburg, and other centers. A world-famous race is the so-called Durban July, which draws a representative field from all over the country. Over 50,000 people attend this event, and to attempt to find accommodations in Durban on the weekend of the July is to seek a needle in a haystack. Auto racing is gaining in popularity, and world-champion drivers have raced in South Africa. Whatever the game, it has its followers in this country. Basketball, badminton, squash, hockey, water polo, yachting, and surfing—South Africans play, and do it well. There is an ice hockey league in Johannesburg and Durban which draws enthusiastic crowds running into the thousands. Visiting teams which come from other countries to teach South African teams a lesson or two often go home a little wiser themselves.

This beautiful country is, in a way, one great playground. There are innumerable homes which have their own floodlit tennis court and swimming pool, and South Africa's country clubs are among the best equipped and most beautiful in the world. Elsewhere in Africa, sport has been one of the major factors in breaking down racial barriers, but it is unlikely to take on this character in South Africa in the near future. African sports facilities are still limited, although they have improved in recent years. There are a number of white sports enthusiasts working with Africans in an effort to improve their playing skill, but the laws of separation which dominate all other aspects of life will prevent interracial competition even when Africans reach the stage where, in the major sports, they could render such competition. As long as there are no non-white members on South African teams traveling overseas, there will be friction at home and questions abroad. South Africans who are asked the obvious question about representation on their teams in the future often resort to an ingenious solution. When, they declare, Africans, Asiatics, and Colored are ready to compete internationally ("and that will be many, many years"), South Africa will be represented not just by one team but by a white team, an African team, an Asiatic team, and a Colored team, all competing separately. In suggesting this, they are completely serious.

FURTHER READING

BROOKES, E. H. *South Africa in a Changing World.* Oxford University Press: Cape Town, 1953.

DRUMMOND, J. *The Black Unicorn*. Gollancz: London, 1959.

KUPER, L. *Passive Resistance in South Africa*. Cape: London, 1956.

MARQUARD, L. *The Peoples and Policies of South Africa*. Oxford University Press: London, 1952.

PATON, A. *Cry, the Beloved Country*. Cape: London, 1948.

——. *Too Late the Phalarope*. New American Library: New York, 1956.

——. *Tales from a Troubled Land*. London, 1961.

Apartheid
and Bantustans

One of the most fantastic experiments ever devised is being tried in South Africa in the 1960's. What was first a general desire for separation and subsequently became a policy of segregation is now reaching its culmination in the application of the philosophy of regional Apartheid. South Africa is being divided into sections, some reserved for Africans, others for whites only. Everywhere, signs are being posted to indicate boundaries. If a white man is entering black man's territory, he needs a permit; and if an African is entering white man's land, he needs a pass. The scheme is beyond belief for many observers, and, more dangerously, many of its details are beyond the comprehension of its supporters. Here is a great, beautiful country, where practically all that has been achieved has been the result of cooperation between white and non-white, being parceled, fragmented into separate units, none of which will have all the grandeur of South Africa and all of which will suffer from isolation and the kind of cramping which in this vast continent of Africa is so rare.

The South Africa of the future, as the Afrikaner nationalists see it, will consist of a number of African areas called *Bantustans*, lying adjacent to regions reserved for whites. Not only will these white and African areas be adjacent, but most of the Bantustans will be surrounded by white land. This is "separate development" at its most complete. Ostensibly, in these Bantustans Africans will not experience the obstacles to progress they experienced in the white man's South Africa, and there will be no limit to what they can achieve. They will, it is stated, have their own universities, own towns and cities, own post offices, own mayors, and will preserve their own customs and traditions. They will, nevertheless, be South Africans. Who will represent them in Parliament? At the Olympic Games? At the United Nations? The answer is always the same: when they reach "the standard of the whites," those will be legitimate questions. Presently, they are not.

South Africans deny that total Apartheid is their goal. Rather, they say, the existence of areas for separate development will aid the African in achieving full citizenship, while not preventing him from entering and working in the white areas. The fact is, Africans will *have* to work in the white areas. In the first place, the land they have been allotted is not adequate for their needs, and in order to pay taxes and perhaps even to subsist, sons must go to the mines and the cities of white South Africa. In addition, the economy of South Africa would come to a rapid stop without the African labor force, and it is not likely that white South Africans want to see every Bantustan completely self-sufficient. Total Apartheid in South Africa is impossible, and Afrikaners as well as English-speaking South Africans realize it. But interracial contact can be limited to that which is absolutely necessary, and this is the national goal of white South Africa. In order to achieve it, the whites are willing to give up their own right of unlimited access to some of the most beautiful parts of the country. They are already putting up with servant problems, because many Africans have been sent away from the cities to their new "national homes," leaving the domestic-servant supply ever smaller. And South Africans are not afraid to spend money and retard the economy to implement the scheme of territorial segregation. It will cost millions and cause some hardship. In order to keep the race pure, the civilization Western, and at least a part of South Africa all-white, no effort is thought too great.

| The Tomlinson Report

At about the same time that the first Nationalist government took office, a group of scientists and researchers was appointed by the Governor-General of South Africa to "conduct an exhaustive inquiry into and to report on a comprehensive scheme for the rehabilitation of the Native Areas with a view to developing within them a social structure in keeping with the culture of the Native and based on effective socio-economic planning." The 9-man group was led by Professor F. R. Tomlinson and hence became known as the Tomlinson Commission. After nearly five years of work, it produced in 1954 the much debated Tomlinson Report, consisting of 51 chapters numbering 3,755 pages, nearly 600 tables, and an atlas of 66 large-scale maps. Thousands of people were interviewed, and the Commission drew upon innumerable officials, institutions, government and university departments, and a large hired staff in its work. The report, when published in abbreviated form in 1955, created a tremendous stir in South Africa. Newspapers serialized it, editors criticized its recommendations, and there were reactions in Parliament. Today, although many of the conclusions of the Commission have not found favor with the government and have been discarded, the report is still one of the most frequently quoted documents in South Africa. Its major findings have certainly influenced the Nationalists in their course of action concerning the Bantustans. As a sample of the thinking or racial matters in South Africa, the Tomlinson Report remains unequaled.

The Commission said that it saw the need for a broad national policy rather than individual tratment for the problems of each African area. It recommended a national policy of separate development of the races, which was

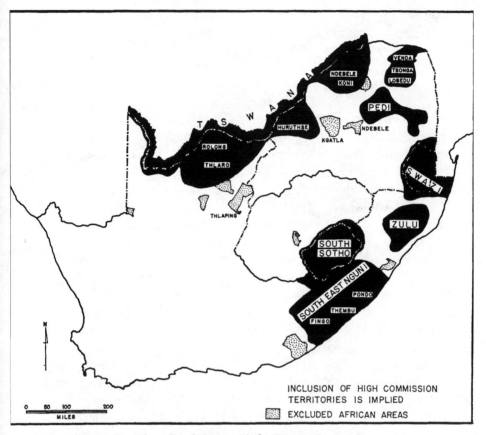

MAP 6 *The consolidation of the Bantu Areas of South Africa into the Bantustans as proposed by the Tomlinson Commission is represented here. Actually, the Bantu Reserves are fragmented into numerous sections, which cover most of the territory occupied by the proposed Bantustans as illustrated. From various points of view, consolidation to the degree envisaged by the Commission was not deemed desirable.*

described as the only way racial conflict could possibly be eliminated. According to the report, there can be no midway between the poles of ultimate total integration and ultimate separate development. Therefore, territories must be established for the Bantu, in which their cultural assets could be preserved and "a basis provided for (African) political development." At the time the Commission commenced its work, there were already 264 scattered African areas in the country. Some of these were very large, such as the Transkei, while others were small and poor, particularly those in the western Transvaal and northern Cape. It was suggested that these areas be consolidated into seven Bantu territories or "heartlands," involving the transfer of some white land to Africans and, here and there, the land of Africans to the white sector. These seven "heartlands" would then become the Bantustans, each for its racial group: there would be a Zulu country, a Sotho country, a Pedi country, etc. An entire new governmental organization was envisaged, and participation of the Africans in the government of their areas was encouraged, but to what limits this political advancement of the African could go was not clearly stated.

The Commission's report has been criticized for allotting too little land to Africans and too much to the whites. This objection is indeed a valid one, but it fails to cut deeply enough. As in other countries, there is good land and bad land in South Africa, and, generally, the land becomes drier to the northwest. It is thus important to note not only how much land the Commission intended to provide for the Bantustans but where precisely this land was located. South Africa, excluding South West Africa, is a country of over 472,000 square miles. Of this, probably less than 20 per cent can by the longest stretch of the imagination be called humid. This amounts to about 94,000 square miles. There are about 10 million Africans in South Africa, and if they were given a proportional share of land, they would get 315,000 square miles of all the land, including 63,000 square miles of the humid lands. The Tomlin-

son Commission proposed to allot to the African population a *total* of 65,000 square miles of all the land in South
Africa, including about 29,000 square miles of the humid
land. If all the Africans in South Africa were to live on the
land the Commission would thus provide, the average density per square mile, in desert and steppe land as well as
humid, would be 154. This would mean 4 acres per person
in a country where even in the Ciskei, one of the better
areas, economists agree that 125 acres is a minimum economic unit! Of course, at no time will all Africans in fact
be in the Bantustans. But even if fifty per cent were at work
in the white areas at any one time, the situation looks little
better. There is no high degree of industrial development
in these areas, and people will for a long time to come have
to live off their land and their animals. It may therefore be
said that land allotment to Africans on the basis of the
Tomlinson Commission's recommendations would sink the
Africans into a battle not for advancement without limits
but for bare survival which would hold back their chances
of improvement in education and other fields as never before.

It would have been surprising for so implausible a
scheme to receive the attention of even the most unyielding
advocates of white supremacy had it not been for an added
factor which made the African land ratio look a little
better. This factor was the possibility of using the remaining British land in South Africa. When the Union came
about in 1910, three territories remained outside its framework. One was Moshesh's Basutoland, which had fought
for its independence and, in a way, succeeded. Another was
Swaziland, which just previous to the establishment of the
Union had asked for British protection and remained a
British protectorate outside the Union in 1910. The third
territory was vast Bechuanaland, which owes so much of
its history to Rhodes. The Commission, in studying the
land situation of South Africa, decided that if any land
were to be given away to the Africans, it should be this land
in the High Commission Territories (as Basutoland,

Swaziland, and Bechuanaland are called), rather than Union land. The combined area of these British territories is 293,420 square miles, and under present British rule, all land in Basutoland and Bechuanaland and about half the land of Swaziland is in African hands already. When the 65,000 square miles in South Africa itself are added to this, some 358,420 square miles (the Commission's arithmetic as published in its Report is in error here) would be available for African occupation. Of the total land in South Africa and the British protectorates, this would be over 46 per cent, a figure constantly quoted by the Commission. It remains far from a proportional segment of the available land.

In "borrowing" the land of the British territories in South Africa, the Tomlinson Commission went beyond its frame of reference and in reality rendered all its allotment proposals useless in practical terms. A number of its suggestions concerning the organization of government in the Bantu areas were of value to the Nationalist government, and the economic considerations were also applicable to substitute schemes. As a practical solution to the "problem" of interracial development, however, the report was not as a whole acceptable. The Tomlinson Commission hoped that the evidence presented would form a justification for South Africa to take over the British protectorates. The Commission requested the government to promise that, if these territories were transferred to South Africa, they would be made national homes for African peoples. The Commission argued that the protectorates already consisted mostly of African land and that Basuto and Swazi not only overflowed from their respective countries but formed a sizeable part of the labor force at work in South Africa's mines. These are valid arguments, but the amount of usable land among the 293,420 square miles of the High Commission Territories is in reality very limited. Bechuanaland lies in the Kalahari Basin, is largely desert and underlain by sandy soils, and its 275,000 square miles provide a subsistence for relatively few people. Basutoland covers only 11,716 square miles, of which a majority lie in high, barren plateaus and steep slopes. Swaziland, best endowed of the three territories, also has a white land-owning population which possesses about half the territory. In the case of Swaziland, the Commission's promise that this would become a national home for the Swazi was somewhat rash. Swaziland's assets are considerable, as is the number of whites profiting from them. Here lies famous Havelock, fourth largest asbestos mine in the world. Here are the world's largest afforestations, the renowned Usutu Forests. There are rich lands, and there is tin, tantalite, baryte, and iron. Had the Tomlinson Commission's report been accepted, this would have become a major crisis area. Many of the people who own land and other rights here happen to be Afrikaners. When conditions in Swaziland are related to the suggestions of the Commission, it must be doubted whether an adequate study was made in this area.

| "Interdependence"

Although only some of the actual recommendations of the Tomlinson Commission were carried out, the general philosophy of regional Apartheid in the Bantustan framework survived and became the foundation for what is presently happening in South Africa. The process of territorial separation has gone vigorously ahead. Without adding significantly to the actual area to be allotted to Africans, the government has started to create a model for the organization it wishes to impose everywhere. Of the 65,000 square miles which the Tomlinson Commission proposed to assign to the Africans, the great bulk was already occupied by Africans anyway, and without committing itself on expansion, the government set about the organization of these areas in the proposed manner. Within these areas, the Bantu are to be given a great deal of what is in South Africa described as "self-government." This means that there will be (under the Ministry of Native Affairs) a permanent Secretary of Native Affairs, who has a number of Under-Secretaries and Native Commissioners, each controlling one region. Each Native Commissioner will be the superior of all Bantu in the territory. Down to this level, all officials are white. Then, the direct administration of the Africans will take place by chiefs, headmen, and their local and advisory councils. These chiefs and headmen are African, and they must answer to the Native Commissioner. This is self-government as seen by the South African government.

Such "self-government" has advanced farthest in the Transkei Territorial Authority (T.T.A.). The Minister of Native Affairs has announced that T.T.A. progress is the reward for cooperation and that self-government will be most rapidly attained in those areas where there is no opposition to the government's plans. In the Transkei, it was not difficult to find people willing to cooperate; there were a number of appointed chiefs and headmen for whom closer contact with the government meant greater security in their positions. Not long ago, the suggestion that such chiefs and headmen should perhaps be elected, not appointed, failed to find favor with the Nationalists. Candidates, it was claimed, would promise all sorts of unattainable goals to their electorate. Thus it was considered unwise, "for the present," to "encourage" such elections. Chiefs and headmen would continue to be appointed.

The Transkei is a large, rather compact, and densely populated area. Though suffering from soil erosion, overgrazing, and occasional droughts, it is among the best land in the hands of Africans in all South Africa. In a way, the government considers the Transkei Territorial Authority the national example of what the Bantustans of the future will look like in the first stages of their development. On paper, with the improvement and advancement of the peoples in the Bantustans, the methods of control of the white "guardian" will fall away. Then, there will be the territorial authorities, which will be "virtually Bantu Parliaments." The greatest strides toward this goal have been taken in the Transkei, where, of course, the Bantu Parliament is still a long way from its formation. Elsewhere, some 450 tribal authorities have been established; these will be grouped first into regional authorities and eventually into territorial authorities. The task was so much easier in the Transkei, because there was already a sizeable region almost entirely in African hands. The government is now faced with the more difficult process of consolidation —a problem which the Tomlinson Commission also faced —of a large number of small and scattered tribal authorities into first a number of regional authorities and then seven or eight territorial authorities. Administrators realize that more than 65,000 square miles will be required, but the expropriation of white man's land to consolidate the tribal authorities is not easy even for a Nationalist government in South Africa. Already, there have been angry scenes at farmers' meetings, when the first maps of proposed consolidated areas appeared. Afrikaners and English-speaking South Africans alike protested against consolidation in many different areas. Even the plans for the final boundaries of the Transkei Territorial Authority faced strong criticism from the farmers of, for instance, the Buffalo River area. These farmers objected not just to the expropriation of land but to the fact that the proposed boundary would come close to their farms. The result would be a great decline in land values, they argued, and this was sufficient reason to keep the boundary far away from such valuable, well-cultivated, and rich land as theirs.

When the whole high philosophy comes down to the simple matter of land and whose it will be, the true color of the Afrikaner and English-speaking South African who talk of their willingness to sacrifice for their country comes out. They are no worse and no better than anyone anywhere in the world who has land and is reluctant to part with it. Some will heed the exhortations of the Nationalists about survival and sacrifices, but most will find some reason for not wishing to sell land for the African authorities. Indeed, what the South African government is actually asking from its white electorate is probably more than any other peacetime government has ever dared to ask, regardless of the urgency of the matter in terms of survival. The South African Nationalist government seems to think that all Afrikaners are constantly imbued with a real sense of guardianship for the black masses and a love for them as children, to be guided toward God and economic advancement. In asking from these people what it does, it forgets that the whites in South Africa are just like people anywhere else— they would prefer to leave sacrifices for later generations, while now enjoying their privileges. This was recognized more clearly in 1956, when the government, reacting to the Tomlinson Commission proposals concerning the consolidation of the African territories, stated officially that "it is, however, unrealistic to indicate at present vague boundaries on maps which involve further European land the acquisition of which cannot possibly now be considered. Future Governments may have reason to return to such theories. At the present moment this is not a practical issue." More recently, it has become a practical issue, a hotly debated one. The tempo of change in South Africa has increased constantly since the Nationalist takeover, and considerable haste is being made in the program of territorial Apartheid. The Nationalist government of 1956 may have been sufficiently patient to wait for "future governments" to solve the matter of land, but the government of today is not.

The government rejected the proposal of the Commission, that white entrepeneurs be admitted into the Bantu areas to promote industrialization. Competition, it was argued, might be damaging to the growth of Bantu enterprises. Actually, the government has its own ideas concerning industrial location in South Africa once the framework of white and black areas has been established. It is obvious that self-sufficiency is not an attainable goal for the Bantustans, and that there is not, among the whites, a great deal of interest in promoting it. The government's attitude regarding industrialization ("will begin with smaller undertakings . . . , service industries in particular should be encouraged") and mining ("premature") in the Bantustans clearly indicates that it envisages other functions for these regions. Indeed, the first improvements are aimed at expansion of agricultural production. Soil erosion is a major problem in many African areas, where overgrazing, forest destruction, poor ploughing methods, and other evils constantly reduce the productivity of the land. Soil simply cannot be left fallow now and then, because there is already a need for greater production, and so, exploited relentlessly, it produces less every year. This is something the government is spending money in combating. As many Africans as possible must be able to wrest a subsistence from their holdings in the Bantustans, and soil care will increase that number.

While whites are not permitted to establish industries within the Bantustans, they are encouraged to locate their enterprises near the borders separating white from black South Africa. Industries requiring large numbers of African laborers, located in suitable white areas near the Bantustans, are "of the utmost importance for the sound socio-economic development of the Bantu areas." These industries are to be run by whites who bring their capital and skills near the borders and draw labor from the great reserves available in the Bantustans. Africans will therefore have to go barely outside their Bantustans to work in these white industries; there will be no excuse for them to travel far into the white man's land. The whole pattern of urban development in the Bantu areas is being directed in such a manner as to facilitate this peripheral industrial development. The government is thus prepared to render the Bantustans and the white areas interdependent. The industries, if ever the labor flow from the Bantustans ceased for one reason or another, would be in the worst possible position. For the Bantustans, the industries are to be a source of employment, a way to relieve population pressure, a manner of making money with which to pay taxes. If urban development in the African areas takes place in premeditated response to this industrial development, as the government intends, then these towns would suffer severely should the arrangement ever fail. These urban centers, whose labor force will undergo a daily exodus to the peripheral industries, would become ghost towns if the borders between black and white South Africa should ever be closed.

It is a fantastic and unique program of planning. Apart from its moral implications, which are the most frequently discussed, there are a number of political, economic, and social aspects which seem more immediately important to the future of South Africa. There is little point in attempting to convince South Africans that the Bantustan concept is morally wrong. Too many South Africans believe the philosophy of Apartheid is the only one which will allow cultures and traditions to come to full fruition. Verwoerd, in a recent speech, drew a parallel with the British Commonwealth, asking whether, if the Commonwealth of nations were in fact one country, the people of Britain would accept the majority rule that would obviously fall to the Indian people by virtue of their numbers. Just as the British would wish to retain their own culture, tradition, and governmental organization, so South Africa's whites now choose to avoid majority rule by those of a lower level of advancement by making separate areas available in which development can take place. Nationalists feel that in so doing they present to the Africans the opportunities which they cannot attain in white South Africa, where they have been deprived of them by law. When questioned about moving the majority of the population, i.e., the Africans, into the smaller and poorer areas of the country, South Africans indicate that they have developed the larger segment they now intend to keep, that they found it empty when they first arrived, that the Africans have no greater claim than have the whites, and that proportional division is a theoretical and impractical solution which would not add in any way to the moral qualities of the scheme.

The real trouble with the Bantustan scheme—even if it were to be given all the time and luck it needs for implementation—is that the economic potential of the Bantustans is insufficient to sustain the kind of development Africans are being promised and that the "interdependence" in the separate development is a one-way dependence. South Africa's white areas could survive without the Bantustans, but the Bantustans cannot survive without white South Africa. Therefore, to see the Bantustans as "little Ghanas" is to mistake the purpose of the plan and the nature of the African areas. It is unbelievably naive to expect that there will really be separate states under the Republic flag, in friendly cooperation. The division of wealth is such as to bring about bitter jealousies, rivalries, and mutual distrust. In so many of its aspects, the Bantustan plan gives evidence of a desire to create what might be called service areas for white South Africa, where large numbers of people, kept busy eking out a subsistence, will not progress very rapidly and in the meantime will remain an easily governed labor supply. There are South Africans who sense the dangers inherent in the plan and object to it on moral grounds and for reasons of self-preservation. They wish to see it abolished or expanded to the point where it appears to have a chance of success. This latter possibility can be disregarded. The amount of land required is so great that it cannot be obtained, in view of the struggles that have already occurred over a few acres. Beside, real self-sufficiency and economic as well as political equality in the Bantustans would mean an economic disaster for white South Africa. Who would work in the Witwatersrand's gold mines? Who would work for low pay in factories and steel mills?

The Bantustan idea by no means implies that white South Africa will be without its African population. What it does mean is that there will be not a permanent but only a temporary, migrant population. An African from the Transkei or some other Authority area will be able to obtain a pass to work in Johannesburg, but after his temporary residence there he must return. Absolutely no non-white will have permanent residence in a white city. Thus there will still be African residential areas, to be sure, but they will be dormitory suburbs for short-term workers. It is a system not altogether different from that which has been in operation for some time, except that the control over illegal entrants into the white cities will be greatly intensified. At any one time, there will be millions of Africans outside their homelands, which will never be capable of containing all their subjects. The result will be a permanent division of the huge African population, increased control over the Africans' movements, and all the advantages of complete domination in the white areas for whites, where Africans are foreigners who can be expelled for improper behavior.

Among the most nebulous features of the Bantustan plan is the political future of the component parts of the newly fragmented South Africa, or, more precisely, the political relationships between these parts. Apart from the Transkei Territorial Authority, the Ciskei, and Zululand, which will have some coastline on the Indian Ocean, the Bantustans are isolated from the outside world and from each other by white land. While a white man can travel from Durban via Johannesburg to Cape Town, a Xosa cannot travel to Zululand without crossing through white territory, for which he needs certain permits. The Bantustans will thus be separated from each other as well as from white South Africa. Some South Africans, fearing that the Bantustans which have a coastline will become danger spots, talk, perhaps somewhat prematurely, of submarines unloading firearms along the marvelous beaches of the southeastern Cape. Others fear that the concentration of Africans in these areas and the exclusion of all whites but a few officials will promote African nationalism, unity, and sense of power. Whatever the case may be, the role of these territories in the governmental framework of the future Republic is not clear. Presumably no South African can visualize his country governed not from Pretoria and Cape Town but from the Transkei and Zululand, but what will, supposing advancement in the Bantustans to satisfy the Nationalists, be the manner of African representation at Pretoria? Will the Bantustans always remain African wards, with internal "self-government" but dominated from white South Africa by a white-elected Parliament? South Africans, who are usually so adept at putting their ideas on paper, have not made any effort to state their intentions here. It would be less difficult to defend the Bantustan plan on moral grounds if it were openly advertised as a temporary phase leading toward eventual equal partnership, but this subject is carefully avoided. The claim would, in any case, not greatly add to the hypocrisy already perpetrated by the planners of territorial Apartheid for South Africa.

The High Commission Territories

When the Tomlinson Commission's report appeared in print, it brought with it a renewal of the discussions, which have now been in progress intermittently for well over half a century, concerning the High Commission Territories—Basutoland, Bechuanaland, and Swaziland. Tomlinson's group wished to incorporate these territories, still governed by the British, into South Africa so that they would become the core areas for several great Bantustans. Their incorporation might have rendered the recommendations of the report somewhat less impractical, but in including them for consideration, the Commission went completely beyond its terms of reference. It was yet another effort on the part of South Africans, however, to effect the elimination of these three countries, which have aptly been described as the geographic anomalies of Southern Africa and which constitute even more of a political liability than does South West Africa's Caprivi Strip.

The three High Commission Territories owe their existence to the political acumen of African leaders in the 19th century. In each country, the people were placed under the

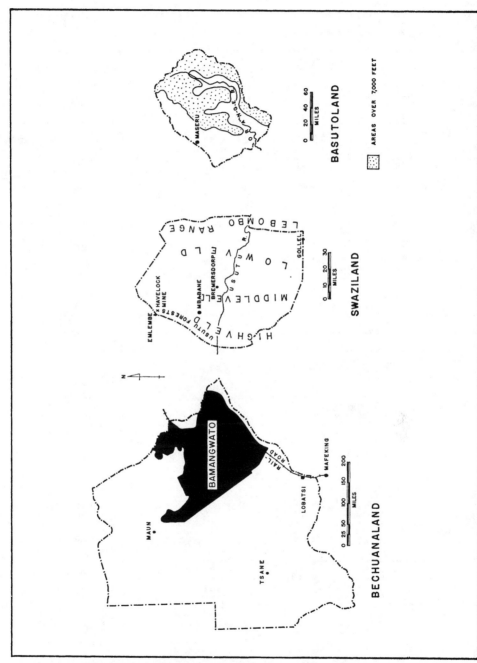

MAP 7 *The High Commission Territories.*

protection of the British while Boer and Briton struggled for supremacy in South Africa. When the Union of South Africa was formed, the High Commission Territories remained outside the new state, although the Act of Union made provision for their eventual incorporation. Today, they are still separate from the Republic, and the Africans in these territories have no wish to join the prosperous giant. In 1909, when the Act of Union was being drafted, the differences between the Union and the High Commission Territories were not so very great; Britain was boss throughout most of Southern Africa, and there seemed no reason to hasten a process that would surely come about naturally in time. It was then difficult to foresee the divergent directions in which South Africa and the High Commission Territories would move socially and politically, although they became evident within a decade after the founding of the Union. The terms of the Act of Union which refer to the territories have frequently been quoted by both proponents and opponents of incorporation, and the present situation is one of those potentially dangerous parts of South Africa's political heritage.

Section 151 of the South Africa Act states that "The King, with the advice of the Privy Council, may, on addresses from the Houses of Parliament of the Union, transfer to the Union the government of any territories . . . belonging to or under the protection of His Majesty, and inhabited wholly or in part by natives. . . ." It states, in addition, that there must be "consultation with the inhabitants before incorporation," and this has been interpreted by the British to mean that the consent of the local inhabitants is required before any transfer can take place. The Union government began to appeal for action on the matter almost immediately after its formation. As early as 1913, Botha wrote to Lord Gladstone, then High Commissioner for South Africa, pressing for the incorporation of Swaziland and Bechuanaland. It was the High Commissioner for South Africa upon whose shoulders the administration of these three territories had rested since

1903. This High Commissioner (whose offices have always been in Pretoria) appoints Resident Commissioners who actually reside in and directly control the territories. Perhaps Gladstone and Botha would have carried their negotiations toward a conclusion had not the upheaval resulting from World War I caused the matter to be shelved. In 1919, however, Botha renewed his efforts to obtain the territories, and he directed himself not to the High Commissioner on this occasion but directly to the Secretary of State, then Lord Milner. An exchange of correspondence followed without any progress being made, and Botha at the end of his period of tenure as head of the Union government was accused of failure in this as in so many other of his efforts. His successor, Hertzog, made strong statements in favor of transfer almost immediately after his election victory, but he was strongly repudiated from London, the Union government being forced to officially deny that Hertzog had expressed any but his own personal beliefs. Africans in Swaziland petitioned against transfer in 1925, but Hertzog, with the assistance of Smuts, continued his barrage of correspondence on the matter. Toward the end of the 1930's, some progress was being made, and an Advisory Conference was held whose members actually visited the territories in order to determine all conditions involved. Before its report could be considered by the British and Union governments, World War II broke out, again interrupting the negotiations at a point where it seemed that action might be taken.

It is important to note that until 1948 negotiations concerning the High Commission Territories were marked by a great deal of cordiality and cooperation and that transfer was seen by many in Britain as an acceptable solution. There was always a sizeable body of opposition, but South Africa did not have the reputation of racial discrimination that it has today, and many felt that racism in South Africa would in time disappear. What was happening in the Rhodesias and Kenya was not basically different from what happened in South Africa, and it is likely that the British government would, without the interruptions mentioned, have transferred the High Commission Territories. What held matters up was the question of administrative complications, plus the small but effective body of opponents who worked hard, in Parliament and elsewhere, against the success of the transfer negotiations. Appeal pamphlets were printed, and newspapers wrote in opposition to the transfer. But those who feel that it was the Africans' own objections which caused Britain to hold the matter in abeyance should note that, as recently as 1953, the united opposition of millions of Africans was overruled in the establishment of the ill-fated Central African Federation (Federation of the Rhodesias and Nyasaland). Until 1948, the Union was rather strongly tied to Britain and the Commonwealth, and it was ruled by a sympathetic government. Transfer was not at all unlikely, and failure was largely accidental.

After 1948, the cordiality which had marked earlier negotiations disappeared. Malan had taken over South Africa for the Nationalists, who were of course the strongest proponents of transfer. However, no formal requests were made until 1952, when the Union Prime Minister raised the matter in the South African Parliament. By 1952, the new trend in South Africa's racial affairs was already well-developed, and in Britain the renewed claim was received with surprise, because it was felt that the policies of the South African government had created a climate distinctly unfavorable to giving up the protectorates at that time. In South Africa itself, the comments made by Malan touched off a vehement dispute. While once again the United Party and the Nationalists more or less agreed that transfer was desirable, the African National Congress indicated its disapproval by stating that it would agitate amongst Africans in all three protectorates to oppose, by force if necessary, the proposed transfer. The A.N.C. had some support amongst liberals and other elements in the South African political scene, and for some time the headlines again concerned the High Commission Territories.

Subsequent developments followed the pattern set in

1952. Malan, in August, 1953, stated that South Africa's patience was coming to an end in this matter. The Union saw with displeasure the progress that was being made toward interracial living in Swaziland and toward constitutional government by Africans in Basutoland. The Union of South Africa, according to its Prime Minister, "could not tolerate the formation of free and independent native states on its borders." The alternative, he suggested, was the creation of core areas of African residence there, in the manner also proposed by the Tomlinson Commission. In 1954, a resolution was passed in House and Senate that negotiations between the two governments be resumed, a suggestion which was rejected by Churchill. Strijdom, succeeding Malan, continued to press for transfer, and he brought the matter up at the 1956 Commonwealth Prime Ministers Conference but met with another refusal. Verwoerd, in the most recent claims to the protectorates, has warned repeatedly that an untenable situation is developing there, and he was particularly explicit on the matter when, during Black Week in 1960, a number of refugees fleeing from the South African police managed to reach Swaziland, Basutoland, and Bechuanaland. Many of the refugees went on to Britain, Ghana, and Nyasaland, but others are still there, a thorn in the flesh of the Nationalists. This event added further significance to the question. If these protectorates can be used as bases for political agitation and perhaps even as strongholds for saboteurs, the South African government cannot be expected to refrain from action.

BASUTOLAND

From the South African point of view, it is not difficult to establish justifications (other than moral) for the proposed transfer of the protectorates. Basutoland, a small entity of 11,716 square miles (about the size of Belgium), lies entirely surrounded by South African territory and has no coast, no major roads, and very few resources. It occupies the highest parts of the great Drakensberg Plateau, and has been called the Switzerland of Southern Africa; it draws thousands of tourists annually. Despite the scenic beauty of Basutoland, however, its soils are of poor quality, agricultural land is scarce, communications are poor, and the potential for economic development is very limited. This protectorate has a population of just under 700,000, including a mere 2,000 whites who are there for purposes of administration and do not own land. Basutoland is one of the most African of African countries, here, amid white South Africa. Wearing cone-shaped hats and beautiful bright-colored blankets, riding donkeys, mules, and horses, the Basuto are among the most picturesque of all Africans. Their country, because of its elevation, can be bitterly cold in winter, and even if there were good soils in these high regions, the growing season would be very short. Geographically, Basutoland occupies a key area, for its deep ravines contain the source of South Africa's greatest river, the Orange. Only in the west is there some lower and less mountainous and less rocky land. Here, and in the lower valleys of the rivers, is where the bulk of the permanent population lives.

It is doubtful whether all 700,000 Basuto could live off Basutoland's limited agricultural land at one time. Soil erosion is such a grave problem that in South Africa soil erosion and Basutoland are practically synonymous terms. Many tens of thousands of young men are always away, working at the mines of the Free State and the Witwatersrand. South Africans declare that without this opportunity for employment, the Basuto would have no way of making a living. The mines, on the other hand, profit from having such a force of good labor in close proximity. A glance at the map shows how much closer this dependable labor source is to these mines than is any other supply of labor in Southern Africa. But one may expect that, if deprived of Basuto labor, the mines could find labor elsewhere, while the Basuto, if deprived of their employment opportunities at the mines and isolated in their territory, would have a difficult time subsisting there.

Basutoland has been plagued with severe droughts and

crop failures, and with each ton of imports, which come through or from South Africa, the dependence of this little landlocked country upon its neighbor is emphasized. Like Africans elsewhere, Basutos eat mainly corn, and this country does not exactly fulfill the requirements for corn cultivation stated previously; hence, corn imports each year are considerable. Wool, mohair, wheat, peas, and a few other exports go to South Africa, but a great deal of subsidization has been necessary. Exploration for diamonds is in progress, and there have been some finds, the significance of which is not yet certain. On the whole, however, the outlook in terms of agricultural and economic development in Basutoland is not favorable. This makes the country virtually defenseless against any South African attempt to force it, through isolation and closing of its boundaries, into submission. South Africa, in reality, does not even have to close the borders but can merely decide that foreigners may not work in South Africa, just as temporary visitors may not work in the United States and in many other countries. This would be the end of Basuto migratory labor and the beginning of great problems in Basutoland. South Africans feel that they are doing Basutoland a continuing favor in prolonging present arrangements, and if Basutoland is to withstand pressure from South Africa, much money will have to be spent on development.

What causes South Africa to cast such covetous eyes upon Basutoland, if the country is an economic liability and possesses little potential? Basutoland is the one country in Southern Africa where Africans have made considerable strides toward self-government. It is a twist of fate that this progress has been made—of all places where it could have occurred—in the one territory that is completely enclosed by the nation in all Africa which objects most strongly to such developments. The British have never controlled Basutoland as tightly as they have the other two High Commission Territories. In 1962, while Swaziland and Bechuanaland are still protectorates, Basutoland is a colony, indicating that it has progressed along the road to-

ward independence. Britain and the Basuto themselves
have always cooperated in an ill-defined manner in the government of this little country. There is a Resident Commissioner, and he has long been advised by the Basutoland
Council, a body of Africans consisting of chiefs and headmen. The Paramount Chief and his subordinate chiefs, at
the same time, have direct power of the Basuto people, and
the link between this power group and the Commissioner
and his council is the bloc of members appointed by the
Paramount Chief to the Basutoland Council.

Political activity in Basutoland goes on at a great rate,
and in view of the fact that there is a high percentage of
literacy (over 90 per cent of the children attend school),
this is not surprising. The Basuto learn avidly and are
willing to make an effort to obtain knowledge. There is
even an institution of higher learning, called the University
College of Roma, not far from the capital, Maseru.
Basuto politicians have been particularly active after 1950,
with the declared goal of a Basuto Legislative Council.
After several setbacks, the Basuto made some spectacular
gains in 1958, when a delegation from the Basutoland
Council went to London to press for the constitutional reform that was taking place in other parts of British
Africa. They achieved a great deal of success, for the
establishment of a Legislative Council was promised,
with a single-roll electorate—a step utterly contrary to
South African practice. Their victory was heralded throughout Africa, especially Southern Africa, and it served to
focus world attention upon the little country. This in itself
was an asset to Basutoland, for the attention the whole
matter received reduced the possibilities for a quiet takeover.

Basutoland's future is by no means secure, and although
it enjoys the asset of progress in constitutional development, it retains its severe geographic liabilities, its vulnerability, its economic problems, and its isolation. Regardless
of the fact that it is surrounded by a hostile South Africa,
Basutoland is landlocked and poor, if beautiful and pictur-

esque. It remains economically dependent no matter what its political status. It is surviving virtually by the grace of South Africa, and if it shows signs of violent political disunity or of intending to support agitators active in South Africa, Basutoland may lose its sovereignty. It will require very mature behavior on the part of a group of leaders and a people who have shown signs of lacking this asset to guide Basutoland through the difficult years to come. Meanwhile, what has been achieved by the Basuto stands as a symbol of defiance for Africans in the other High Commission Territories, who, similarly threatened with incorporation by South Africa, intend also to resist.

BECHUANALAND

Bechuanaland, in many ways, is the exact opposite of Basutoland. It is a vast, flat country, underlain by sandy soils which form drifting, windswept plains here and swamps there, while nowhere supporting intensive agriculture. Somewhat more than 300,000 people live scattered over 275,000 square miles of desert, bush, steppe, and morass. About the size of Texas, Bechuanaland is hardly able to sustain a few hundred thousand wandering cattle-herders. Flat and monotonously unending, instead of mountainous and small, Bechuanaland has one asset which Basutoland has not: a boundary with the Central African Federation. Bechuanaland comes close to being surrounded by South Africa and its possession, South West Africa: the Caprivi Strip and the northern Transvaal miss by only a few hundred miles pinching the territory off. It occupies the Kalahari core of Southern Africa, and again the South Africans point to the geographic homogeneity of the area now politically divided, but their claim is of less substance than that affecting Basutoland. Though landlocked, Bechuanaland is not isolated by one nation from the outside world.

When Rhodes made known his intention to place Bechuanaland under the administration of the British South Africa Company, the Bechuana (Tswana) chiefs went to London to request Imperial protection. Eventually, an

agreement was reached whereby the Bechuana would indeed have a British protectorate and would not be under
the direct rule of the Company, while in return they promised to assist in the building of the railroad across their
land to the Rhodesias. In 1962, there still is no direct rail
link between the Transvaal and Southern Rhodesia. The
Bechuanaland railroad is the only connection (with the
Johannesburg–Lourenço Marques–Salisbury rail link)
between south and north in Southern Africa. Bechuanaland
is still a protectorate, and little progress has been made in
the improvement of education and health in the territory,
which remains poor and backward. Apart from the railroad, communications are practically non-existent. The
mode of life of the semi-nomadic Bechuana cattlemen has
changed little, and this is the country where some of the
Bushman clans still survive. Tsetse fly, malaria mosquito,
yellow fever, bilharzia, and other pests and diseases have
not yet been eradicated. Water supply is problematic, rainfall reliability is low, drought common, and cattle frequently die from thirst. A few thousand whites are engaged
in administering the country, and a geologic survey is mapping the area and prospecting for minerals. There is a
little irrigation, but cattle remain the sole export. A few
mission stations make some medical facilities available, but
social amenities are virtually absent.

Like Basutoland, Bechuanaland is ruled from London
via a Resident Commissioner and his administrative officials. The Bechuana, however, have far less power in their
country than do the Basuto in Basutoland. The Commissioner meets with an Advisory Council, but though many
chiefs have indicated their desire for a Legislative Council
in Bechuanaland, and a legislative body was petitioned for
by Africans as well as whites in 1958, there has been little
progress toward establishing one. The whole matter of
economic and political advancement in Bechuanaland was
done much harm by the marriage of the chief of the Bamangwato people, Seretse Khama, to a white woman in
1948, while he was studying in Britain. Actually, it was not

the marriage itself which caused the trouble but the reaction of a section of the Bamangwato and the behavior of the British government in the matter. In Seretse's absence, a Regent had been ruling the Bamangwato people, and this Regent objected to the mixed marriage of the chief, soon to return. He set about preventing the Bamangwato from accepting Seretse and his white wife as their chief and queen. In Bechuanaland, as elsewhere, the chiefs and headmen are left more or less to themselves in ruling the African peoples, and problems of a local nature are solved by Africans. There were several mass meetings, at which the Bamangwato displayed deep division and resentment, but in the final meeting concerning this question, it was clear that the majority wished Seretse to return and would accept his white wife. When the Regent then left the tribal territory and further dissention occurred, the British intervened, prevented Seretse from returning, and prevented the Regent from returning also. Eventually, after years of negotiations, Seretse Khama returned to Bechuanaland, there to be received by a show of allegiance such as had never before been witnessed. But by then it was 1956, and progress had been halted since the marriage in 1948. The Bamangwato, whose land, in terms of resources, may be the best in Bechuanaland, had been unable to decide without their chief on matters which greatly affected their advancement. Thus mining companies which wished to explore and purchase rights were kept waiting, no schools could be constructed, and there was general stagnation, much to the detriment not only of the Bamangwato but of all Bechuanaland.

In the eyes of Africa, Britain's failure to abide by its own established procedures for the government of its dependencies was evidence of its insincerity. Indeed, it was unfortunate that this marriage between a black chief and a white woman led to instability in Bechuanaland and that British intervention was the ultimate result. It seemed clear that Britain shared South Africa's philosophies opposing mixed marriages and that an effort was being made to keep this particular mixed marriage out of Southern Africa. It also appeared that the tradition of non-interference in tribal affairs, generally preserved by Britain, was needlessly broken. In the event, Bechuanaland received world attention, but progress, such as it had been prior to this crisis, came to a standstill.

South Africa's claims to Bechuanaland as being a necessary part of the Bantustan plan are not well founded. Although very large, Bechuanaland is unable even to provide a reasonable living for 300,000 people, and while its area would add significantly to the total square mileage the South Africans propose to allot to the Bantu, it adds little to the usable land able to sustain the millions of people involved. No doubt progress will reach even Bechuanaland. But the funds required to raise the standard of living in Bechuanaland are far in excess of the total appropriated for the entire Bantustan plan. By itself in its present state Bechuanaland cannot make a contribution to any plan, certainly not the Bantustan scheme. The South Africans also may be mistaken if they believe they will find the Bechuana ready to submit to their rule. Events between 1948 and 1956 have shown the local Africans the undesirability of external rule, and claims for the establishment of a Legislative Council are being renewed. Political consciousness is by no means as universal as it is in Basutoland, and both education and literacy lag far behind, but Bechuanaland is changing. With the rebirth of the South African Republic, the British decided to change the site of the Bechuanaland capital—which, anomalously, lay outside Bechuanaland at Mafeking in the northern Cape—to a place within the territory. The site of the new capital (and there are not many to choose from) will probably be Lobatsi, where several government offices are already located. This and the protection afforded to fleeing refugees at the time of Sharpeville in 1960, give evidence of the fact that Bechuanaland intends to retain its identity as a separate political unit and does not jump to South African orders. In order to survive, it will need millions. Its major market for the cattle exports is South Africa. Those Bechuana who work outside their country work mainly in the Republic. What little economic development there is remains dependent upon the whims of South Africa, which could not throttle the territory but which could cause chaos by closing the boundaries. Bechuanaland must be prepared to absorb the effects of this action. This is why economic development in the High Commission Territories, including oft-forgotten Bechuanaland, is of the utmost importance and a matter of great urgency.

SWAZILAND

Among the three territories under the High Commissioner's administration, Swaziland is in many ways unique. It is by far the smallest, its 6,704 square miles equaling just over half the area of tiny Basutoland and less than one-fortieth of that of Bechuanaland. It also has the smallest African population (240,000) and the largest white population (7,000). Swaziland lies in the southeastern corner of the Transvaal, but although it is landlocked, the Republic does not surround the territory. It has a short and vitally important boundary with Moçambique, the Portuguese province in southeast Africa. The protectorate extends in part over the highveld and in part over the middleveld and lowveld and is therefore very diversified in spite of its small area. Its beauty is unequaled in the subcontinent, with high, misty mountains descending onto forest-covered plateaus, which in turn lead down into the grassy plains of the lowveld. Great rivers rise on the highveld and cross Swaziland before entering adjacent Moçambique or northern Natal, and herds of impala remain in the more remote lower areas. Abundant rain falls in the higher parts of the country, but the lowveld is dry, requiring irrigation to make the good soils produce. Swaziland is a land of opportunities, and were it not for its political situation, a bright future might be prophesied.

Like Bechuanaland, Swaziland is a protectorate, but more so by accident than by negotiation. The Swazi deputation which went to London in 1893 requesting protection met with refusal, and the British were quite prepared to let the Transvaal Republic take the territory over. From 1894, the Afrikaners held sway over the country, and in all probability Swaziland would have lost its identify altogether had it not been for the Boer War. With the defeat of the Transvaal, Swaziland finally received British protection in 1903. This, however, did not mean that the problems created previous to the war were now eliminated. Careless concessions had robbed the Swazi of most of their land, and one of the first problems the new British administration had to face was the relocation of Swazi who were still on land which no longer belonged to them and the setting up of a mechanism by which they could regain some of this land for their own use.

A special commission was established for these purposes, and its first action was to deduct one-third of the land from all concessions and return this land to the Swazi. This provided the Swazi with some of the land they badly needed, but the Swazi themselves, in return, had to agree to leave land which was not theirs and settle elsewhere unless a private agreement with the landowners could be reached. They were given a 5-year period, from July 1, 1909, until June 30, 1914, during which they could not be compelled to move. After the date of June 30, 1914, however, they would either have to leave for the Swazi areas or stay with permission and by special arrangement. No great Swazi migration resulted from these decisions, for the great majority of the Africans managed to make satisfactory settlements.

The land question in Swaziland is still not settled, and it is the one High Commission Territory which still has many white landowners. This has had favorable and unfavorable consequences. Because of the fact that white and African farms lie intermixed, there has over the years resulted a great deal of interracial cooperation. A drought hits white and black farmer alike, and there is nothing like combating a common enemy to draw people together. To see Africans and whites work together, using each others tractors, aiding each other in terracing the land, and making similar profits based only upon the market price and not upon the skin color of the farmer, is to see hope in Southern Africa. There are African farmers—though admittedly not many—who earn as much as do wealthy farmers anywhere. It is the existence of such a group that is so important, in that it shows the less fortunate Swazi what can be achieved, while removing a cause for racial friction. In a way, Swaziland is a showpiece for all Africa.

The problems arising out of this arrangement include those of administration, which because of the scattering of the Swazi lands is not easy, and penetration by alien whites, which is much the more serious. The fact that there are no "white lands" in Basutoland and Bechuanaland means that there is no avenue of penetration there, that no sizeable white population can be established and in time make its political influence felt. Swaziland does have the liability of white penetration. An increasing number of Afrikaans farmers are purchasing land in Swaziland and either running their farms from South Africa or living in the territory. The former method of farming—absentee ownership—is obviously undesirable from the point of view of the territory. Those whites living in the protectorate, however, are exerting an ever stronger influence in the political affairs of the little country. There is an increasing number of Afrikaners among them, and today some sixty per cent of all whites living in Swaziland were born in South Africa. Particularly in the south, this Afrikaans element is strong. Here, farmers' meetings are already sometimes run in Afrikaans and not English, which is the official language here. This is a source for great concern on the part of the administration. The Afrikaners have financial support from South Africa and do not have to renounce South African citizenship in order to live in Swaziland. There have been repeated calls for the closing of the boundaries and the termination of the present free-access arrangements, so that this tide of Afrikaner penetration can be stemmed. At

present, there are no border formalities anywhere between
South Africa and the High Commission Territories, and
the cost of establishing border posts would be prohibitive.
The penalty now paid for this lack of control is heavy. The
Afrikaner voice in Swaziland is growing ever stronger, and
Afrikaners possess some of Swaziland's best farms.

Indirectly, this Afrikaner settlement in Swaziland has
other serious consequences. When the land was divided
between Swazi and whites many years ago, it was made
clear to the Swazi that, if they would save money for this
purpose, they would be free to purchase back any land
which whites might be willing to sell to them. Since the
early days, the Swazi have saved religiously, and they have
already bought just short of 200,000 acres, which, at the
white man's prices, is a real achievement. Today, the Swazi
again hold more land than do the whites, and they hope in
time to be able to regain their territory in its entirety—although some Swazi also see the value of the interacial cooperation existing at present. Clearly, the Afrikaners who
have bought land in Swaziland will never relinquish it to
the Swazi, and it can be stated that land thus sold is lost
to the Africans, which leads to resentment. The Swazi blame
the white administrators for not preventing this immigration of foreigners, and it is difficult to find an answer to
the charge. Meanwhile, the process continues, and as more
English people leave because they foresee trouble in the
country, Afrikaners take over. Today Afrikaners have a
major stake in the southern half of the protectorate.

This is one of the reasons for the allegation that the
Tomlinson Commission was extremely rash in stating that
the High Commission Territories would be made the national homes for Africans. The Tomlinson report was, in
part, and effort to woo the Swazi into support for the
scheme, for the Swazi resent not having more land in their
hands. But it must be seriously doubted whether several
thousand whites could be moved from Swaziland, in view
of the fact that the whites, including many Afrikaners, hold
excellent agricultural land, have a stake in the irrigation

scheme of the lowveld on the Great Usutu River, own the mines, and dominate the economic life of the territory. Those who know Swaziland and its white inhabitants will deny that the Tomlinson Commission's plans were within the realm of possibility.

In terms of resources, Swaziland is better endowed than the two other High Commission Territories put together. The highveld section of Swaziland, in the west, is composed of mountains reaching their culmination in Emlembe, over 6,100 feet in height. Here are great granite domes which wrest as much as 70 inches of rain from the moist winds rising against the plateau edge. In 1955 in the Usutu Forests, which lie here in the moist west, some 120 inches of rain were recorded, but the average for this whole area is between 50 and 60 inches. For Southern Africa, this is a high total, and Swaziland is indeed fortunate in this respect. From the highveld and forested plateaus, the bush-clad land drops eastward to the lowveld, a trough only a few hundred feet above sea level, yet not a part of the coastal plain. This lowveld is separated from the coastal plain of Moçambique by the Lebombo Range, a north-south trending mountain chain through which the rivers crossing Swaziland have cut deep gorges. Along this range runs Swaziland's eastern boundary. The lowveld itself is dry, partly steppe, but fairly grassy. In this area, droughts are frequent, but this is cattle country, these disasters notwithstanding. In the south, the mighty Usutu River meanders across the area, and here a great irrigation project is now being carried out. Swaziland is beginning to export sugar, grown here at Big Bend, and the expansion of this project will greatly increase the territory's cash-crop production.

One of the great problems in Africa is the teaching of soil conservation to Africans who have never taken much care of their land. Overgrazing (there are thousands upon thousands of goats in Swaziland) and bad farming methods have here, as elsewhere, caused the destruction of much valuable soil. At Bremersdorp is the Swaziland Agricultural Station, where a group of dedicated men are engaged in mapping, classifying, and determining the yield capacity of the country's soils. Groups of Africans are being trained to teach others how to protect the soil, contour farming is promoted, and crop suitability is studied with reference to the available soils, which in parts of Swaziland are quite good. Prominent among the men working on these vitally important matters is George Murdoch, an officer in the service of the Agricultural Station for a number of years. Murdoch personifies the kind of worker for Africa who is never heralded, whose successes go unsung, whose labors go poorly rewarded, but whose activities are among the most important and beneficial to Africa's people. Out in the field, uncounted hours are spent in the blazing sun, mapping soils in poorly charted territory, fighting off mosquitos, drinking filtered water, living in leaky tents. From sunrise to sundown, Murdoch and his contemporaries are on the go, performing far beyond the call of duty and displaying the kind of affection which is of so much more value to Africa than are the dollars of wealthy tourists and the bubbling words of negrophiles. In Swaziland, as in so many other countries of Africa, the results of this work have had a profound effect upon progress and prosperity. Everywhere, contour ploughing gives the countryside a neat and cared-for appearance. New crops, introduced as a consequence of research carried out by agricultural experts, are making money for white and non-white farmer alike. Proper fertilization methods are becoming known. Gulleys caused by man-induced erosion are closed and planted with grass. In Swaziland, there is still hope that soil erosion can be conquered, and it is certain that Africans are taking to better farming methods as taught by the selfless agricultural workers. The aspect of this territory improves constantly—a monument to the labors of a few dedicated men and a compliment to farmers of all kinds.

Among Swaziland's great assets are the Usutu Forests, said to be the most extensive single area of afforestation in the world and located in the higher, moist regions of the territory. There are well over 150,000 acres of pine forests, owned by several corporations, private as well as government. With the forests reaching maturity, wood and wood products figure increasingly among the country's exports. In addition to the acres under pine, there are also wattle plantations, and bark from this tree is also exported. The indigenous forests, which are particularly dense in the gorges in the hills and mountains, are not much used except by small furniture manufacturers and by farmers for fencing.

As elsewhere in Southern Africa, corn is by far the most important food crop in Swaziland, and sorghum is second. A little wheat and rice are cultivated by those farmers who can afford to irrigate their land. Swaziland, in some ways, is an extension of the prosperous eastern Transvaal, and many of the crops which do well near Nelspruit and Barberton also thrive in the protectorate. Citrus fruits, pineapples, and the like are exported, and sugar production is rapidly increasing in the irrigated lowveld. Among the other crops, tobacco and cotton have proved their suitability in the territory, and there can be no doubt that Swaziland's agricultural potential is still considerable, particularly with further irrigation and improvment in farming methods. Compared to Bechuanaland and Basutoland, Swaziland presents a favorable picture in terms of agriculture. Well over 75 per cent of all the cattle in Swaziland are owned by Africans, and dipping practices have been introduced. There was some opposition to this, but in general the results have been an improvement in the strain and an increase in the numbers which have withstood disease. In fact, the number of animals in the hands of the Swazi has grown so much since 1910 that the purchasing of land was more than a matter of desire; it was an economic necessity, as it still is today. The Africans, unfortunately, do not make proper use of fencing to prevent overgrazing. As soon as land has been bought from whites, the fences are more often than not torn down, to demonstrate that it is communal property. Unfortunately, this practice has prevented the best grazing methods from being carried out, and as a consequence the land has been damaged. However, the Swazi are now learning good grazing methods as well as soil conservation.

Swaziland's economic resources are not limited to soils and vegetation. What drew the large number of white concession-hunters and settlers in the latter part of the 19th century, to form the nucleus of the sizeable white population of today, were the rewards in prospecting. As in the eastern Transvaal, gold was found and mined in Swaziland's northwest. Deserted shaft-heads, decayed and overgrown by bush, attest silently to those days of frantic activity. Occasionally, fortune-hunters descend into these old workings to extract yet another few ounces of gold, but organized gold mining does not take place any longer. Since the early days, other minerals have become much more significant than gold itself ever was. Today asbestos produces about 70 per cent of the income the territory derived from exports. There is a tremendous chrysotile deposit in the northwest, which is among the four most important in the world, and Havelock mine regularly ranks fourth in terms of asbestos production among the world's mines. In addition, tin has been found and is exported in small quantities. The discovery of tin has not been an unmixed blessing for the protectorate. The private companies which extract it use water blasting methods to loosen the gravel, and enormous scars remain in the countryside where mining has taken place. These scars have initiated more severe erosion gulleys which form a problem for the soil-conservation experts. In addition to tin, baryte is mined in small quantity, as is tantalite. Needless to say, all this mining is in the hands of white people, who have provided the capital. The Tomlinson Commission, when it suggested Swaziland as an African territory, failed to take these matters into account. It is difficult to visualize the inducements which would have been needed to cause the whites to leave this territory. If arrangements had been made to keep the whites at Havelock and other mines, then the promises of the Commission would have been diluted and the Africans betrayed.

There are also mineral deposits which have only been located, not yet exploited. Near the capital of the territory, which lies in the high west, a top-grade iron deposit has been reported, and in the dry lowveld, prospects are enhanced by a coal reserve. Radioactive minerals have been found and are presently being investigated. For a territory as small as Swaziland, there is much justification for optimism as far as its chances for economic development are concerned. Mineral deposits, rivers affording excellent sites for dam construction, irrigable lands, workable soils, moist regions, and climatic diversification make a favorable impression. There are some food-processing industries, and an improved road system is under construction.

What, then, has held Swaziland back to cause it today to be economically dependent upon the Republic of South Africa, and what has delayed the exploitation of all these resources in a subcontinent where exploitation has always been of the order of the day? There are several reasons, and together they constitute Swaziland's major liabilities. They are all artificial, and their significance increases each year. In 1910, when Swaziland remained outside the Union, an agreement was signed, referred to as the Customs Union, which stipulates that all exports of Swaziland must travel through South Africa before eventual overseas shipment. This is one reason why there is still no rail link between the natural port for Swaziland, Lourenço Marques, and the protectorate. The British once signed an agreement with the Portuguese that a railroad would be constructed from Lourenço Marques through one of the gorges in the Lebombo Range and into Swaziland. The Portuguese, eager to promote trade through their city, built the railroad on their side, but the British never carried out their part of the agreement, which has given rise to some resentment among Portuguese. More important than Portuguese resentment has been the effect for Swaziland. The Customs Union and the absence of an eastern railroad—which because of the Union would be useless even if constructed—have caused the territory throughout its history to look to the west and not the east. Although there is a short border with Moçambique, this stretch of contact with the outside world has not yet attained much significance. Swaziland is entirely under the domination of South Africa. It uses South African currency. Its products go to or through the Republic, which is the market for most of them. The Republic determines what shall be exported and not so long ago restricted the export of cattle to such a point that this export was virtually eliminated. South Africa operates the communications services inside the territory. There are still no railroads in any part of Swaziland, and buses of the South African Railroads connect points within the protectorate with railheads outside its borders. This means that all effective communications are in the hands of South Africans, who of course have a stake in controlling this vital aspect of the economic life of the territory. It also means that the exploitation of such deposits as iron ore and coal are delayed; no private company, for example, would want to invest in an area where there is no means of transportation other than trucks and buses over roads which are mud tracks part of the year. The government of Swaziland does not have the money, and the South Africans are well served by the delay. The result? Economic stagnation and continuing vulnerability.

If the Customs Union with South Africa were repealed, Swaziland would no doubt suffer severely, since it also benefits from this Union in terms of assured marketing of products the Republic wishes to import. Talk of termination of the agreement is premature. Although Swaziland's resource base is better than that of Basutoland or Swaziland, it is no less vulnerable to South African economic aggression. In a way, the position of this little country is sadder than that of its larger sister territories. The chances being lost in Swaziland are so much more easy to grasp, or

would have been. It is not surrounded by the Republic, nor
is it so far inland as to be isolated by sheer mileage. Swaziland is within 50 miles of a large ocean port which is more
than able to handle its exports. Swaziland would not be a
rich country, but with some efforts the territory could have
developed a degree of economic independence from South
Africa.

One reason for the ease with which the South Africans
continue to dominate the economic life of the protectorate
is migratory labor. Like the Basuto, thousands of Swazi
stream to the mines of the Republic each year, there to
work, save, and return with money for taxes and other
things. Swaziland would be hard hit by a termination of
this arrangement. In addition, the Swazi, nearly a quarter
of a million of whom live in Swaziland, also occupy large
parts of the southeastern Transvaal, where their numbers
are variously estimated at somewhere near 70,000. They
benefit from the present border arrangement, for they
travel into and out of Swaziland in large numbers, particularly at the time of the Ceremony of the First Fruits, when
tens of thousands of Swazi gather for century-old rites.
Swaziland is one of the very few political units in Africa
occupied dominantly by only one indigenous people. The
area which is today Swaziland has since the arrival of the
first Swazi—perhaps as early as 1750 under Chief Ngwane
III—been occupied by the Swazi nation; not even the Zulu
succeeded in dislodging them. But the Swaziland of old was
much larger than the Swaziland of today, and many Swazi
still live outside their national territory. Many of these
still come annually across the border into Swaziland to pay
homage to their chief, Sobhuza II, to whom much of the
credit for what has been achieved in this small country must
go. Thus a number of factors complicate the question of
Swaziland's future as a political entity. There is a young
generation of Swazi who say they will resist until death the
incorporation of their country into the Republic. There are
many Swazi to whom the present arrangement of migratory labor and open borders is a matter of need. There are

whites who sympathize which the South African government and who would like the African in Swaziland "put in
his place" as he is in the Republic. There are other whites,
the British, who would like to see the borders closed and
the soil wiped off the Union Jack which still waves at
Mbabane.

The central figure of the Swazi nation is its Paramount
Chief, Sobhuza II. Probably the most able and certainly the
most respected chief in Swazi history, Sobhuza has proved
himself a clever politician and, incidentally, a good busi-
nessman; he has a surprising number of interests in such
enterprises as hotels in the area. Sobhuza and his two
Councils maintain a traditional rule over the Swazi, while
the Resident Commissioner deals with other matters. The
Resident Commissioner is guided by an Advisory Council,
initiated in 1921 and consisting of whites, in his administration of the white population of the territory. Thus the
familiar pattern of dual rule is repeated, and the present
arrangement was established by proclamation as recently
as 1950. Of the two Councils over which Sobhuza presides,
the lower, the popularly elected Libandhla, is increasing its
power, while the other, formerly more important and
consisting of a group of prominent Swazi, is declining in
significance. The one Swazi political organization of any
consequence is the Swazi Progressive Association, of which
many of the younger Swazi are members and many of
whose adherents are in the Libandhla Council. For matters
which require consultation, the Resident Commissioner
meets with the entire African Advisory (combined) Council, at which time Sobhuza attends. Although things are
changing in Swaziland as elsewhere, the Paramount Chief
still does not make many decisions without consulting his
mother, who is called upon to reason with the Chief if
members of his Council fail to do so. In a way, the Swazi
possess a dual monarchy. The mother of the Paramount
Chief has under her care such important matters as the
rain-making materials, and in the life of the Swazi she still
plays a very important role.

There is a danger of white and African political organizations drifting apart in Swaziland. To counter this danger, whites as well as Africans have called for a revision of the system which now allows them to exchange views officially only at occasional meetings between the Resident Commissioner and African representatives. There is also a small Colored community in the protectorate which desires representation and which has displayed some restlessness. Since no law prohibits mixed marriages, a number of Colored people have made their way to Swaziland from South Africa, and their influence is becoming noticeable. But although there may be misunderstanding in some fields, there is a refreshing spirit of mutual confidence and cooperation between a large number of the white people and many of the Africans. There is no official racial segregation, although in practice it does occasionally occur, particularly where there are many Afrikaners. Frequently, on the other hand, interracial parties and meetings are held, and without any self-consciousness. In Mbabane, the capital, particularly, the atmosphere of non-racialism is a pleasant surprise, proving again that it is the absence of laws preventing social interracial mixing which most promotes moderation on all sides. Afrikaners who reside in the territory say that they find no need for what is to them repulsive interracial contact—they can go their own way without having to make any more such contact than they ever did in South Africa. Swazilanders (white residents of Swaziland, as distinct from Swazi, the African variety) prefer to have the opportunity for such interracial mixing, without making a fetish of it. It would not be correct to suggest that the whites in Swaziland (other than South Africans) have all the virtues and none of the shortcomings of whites in other British territories, but they have an undeniably sound attitude to life in the African territory. Many Swazi who have been to work in South Africa return with a new appreciation for the harmony which prevails in their homeland, although they also frequently express disappointment at the lack of opportunity, of which they saw more in South Africa. They would like more schools and better hospitals, and they blame the local whites for the economic problems of their country; but Swazilanders as well as Swazi appear to treat these things as "our" problems, and blame is not cast only across racial lines, as it is most commonly in South Africa.

Swaziland's capacity to resist absorption by South Africa will, unquestionably, depend upon its degree of economic development. At present, neither Basutoland nor Bechuanaland appear to have the assets required to greatly accelerate this process, but Swaziland is more fortunate. True, African education in Swaziland still lags far behind the desired pace of improvement, but the Swazi are very interested in learning and are voluntarily contributing to a Swazi Education Fund, through which three Swazi National Schools are now in part maintained. Health conditions must be bettered, but the present situation is no worse than it is in many other parts of the subcontinent. Swaziland, moreover, has a resource base which constitutes its great hope for the future. There is an adequate labor source which today still finds much of its employment outside the boundaries of the territory. The country is physically diversified, and though small, it has more of the requirements for independence than have many other, less endowed territories. Size alone is no criterion upon which to base estimates concerning the chances for successful self-government. It is much larger than Lebanon, exceeds Jamaica in size, and is only slightly smaller than El Salvador or Israel. Among the states of the world, Swaziland would by no means be the tiniest.

| Deciding the Future

What will determine more than anything else the course of Swaziland's future is the manner in which the territory's potential is realized. It is unfortunately true that Swaziland has all too often been bypassed when money was spent on the High Commission Territories; Basutoland seems long to have been London's first love. But while Basutoland can really only survive by the grace of South Africa, Swaziland could exist without the South Africans' blessings. Only economic strength will give Swaziland the fortitude it may need, and the ingredients for that strength are present. Many whites and most Swazi recognize this fact and are willing to work for the improvement of their country's condition, but money and outside assistance are required. Rather than demand further political advancement, the inhabitants of Swaziland will serve their country best if given the opportunity to accelerate the development of the economy. No spectactular improvements can be expected, for Africa resists rapid change. There must, on the other hand, be no stagnation. South Africa has a very strong point when it states that the territories would benefit materially if taken over by the Republic, and it is a tragedy that this should be the case, particularly in Swaziland. The British have neglected the High Commission Territories, in part because of the diplomatic troubles that could have arisen had there been a great deal of advancement. There was, perhaps, some cause for relative neglect of the territories while the Union of South Africa was still in the Commonwealth. Now that the Republic has withdrawn, however, Britain may take a stronger and more independent policy concerning the High Commission Territories. Thanks to the competence and interest of several administrators, as well as the sensible attitude of a part of the white population in all three territories, there is already a great deal of interracial cooperation and understanding. Everywhere, problems will arise, and in Bechuanaland and Basutoland, the chances for success are not very good. Swaziland, however, could turn to the east and establish trade communications through Lourenço Marques. It could—not in 1962, but on the basis of the available resources more fully exploited—survive termination of the Customs Union with South Africa. If the High Commission Territories are absorbed in the near future, it will be a tragedy. It will be a double tragedy in the case of Swaziland.

South Africa's own interests may, at the moment, not be served by renewed efforts to obtain the protectorates. Driven into isolation by its withdrawal from the Commonwealth and by its attitude on South West Africa, which has led to severe censure and defeat in the United Nations, the Republic may let matters which can heap further criticism upon its shoulders rest for some time, until the political climate has changed. Effectively, the Republic dominates the territories anyway, and aside from the annoyance they give by providing political refuge, they do not endanger the Republic's national security. Among Afrikaners, the attitude expressed by Malan, that the Republic "cannot tolerate the development of free and independent African States on its borders," still prevails, and a delay in action cannot be taken as a cessation of the claim to the protectorates. South Africa still maintains so much control in the territories that they cannot be seen as actual or even potential centers for anti-South African action of a nature that would precipitate Republican intervention. Some Basuto went to Moscow and attended Communist youth congresses, and no doubt there are people with Communist sympathies in the territories, but it is difficult to visualize the High Commission Territories as centers for really effective political activity. Political activity may increase radically if the whole situation in the buffer zone to the north changes. But until then, it is unlikely that South Africa will present any additional reasons for the incorporation of the protectorates. While there are no new foundations for its claims, South Africa may decline to act in the matter. It is this period which the High Commission Territories must use in order to advance economically faster than ever before. No one wants to perpetuate the political fragmentation of the continent of Africa, but neither should people be compelled to subject themselves, involuntarily, to a political and social system which may bring material advantages but which is alien and oppressive. If it is true, as the Afrikaners so often claim, that Africans "do not know the difference," then any move should be delayed until they do. In a continent where so many chances have been lost, this is one case where a repetition must not occur. The whole question need have nothing to do with the moral issue of South Africa's Apartheid laws. It is simply a matter of whether people who have for a century fought and negotiated for their land and for a degree of self-determination shall be allowed to decide for themselves what their political future in fast-changing Africa will be.

FURTHER READING

ASHTON, E. H. *The Basuto.* Oxford University Press: London, 1952.

GREAVES, B. *The High Commission Territories.* Edinburgh House Press: Edinburgh, 1954.

KUPER, H. *The Uniform of Colour: a Study of White-Black Relationships in Swaziland.* Witwatersrand University Press: Johannesburg, 1948.

MASON, P. *An Essay on Racial Tension.* Oxford University Press: New York, 1954.

ORCHARD, R. K. *The High Commission Territories of South Africa.* World Dominion Press: London, 1951.

PIENAAR, S. AND SAMPSON, A. *South Africa: Two Views of Separate Development.* Oxford University Press: London, 1960.

REDFERN, J. *Ruth and Seretse: a Very Disreputable Transaction.* Gollancz: London, 1955.

SITHOLE, N. *African Nationalism.* Oxford University Press: London, 1959.

TOMLINSON COMMISSION. *Summary of the Report.* Government Printer: Pretoria, 1955.

Forced Wedding: Federation in the Rhodesias and Nyasaland

The Republic of South Africa and the territories it rules and dominates constitute white Africa. Separating this stronghold of absolute white minority control from the Africa now under African control are five political entities. These five—the two Portuguese territories, the two Rhodesias and Nyasaland—form today's buffer zone, separating white from black Africa while possessing some of the characteristics of both. White Africa is represented in the buffer zone by a modification of the segregationist rule of the south. In the Rhodesias, particularly Southern Rhodesia, and in Angola and Moçambique in a different manner, the white man rules; but the voice of the African is heard louder and clearer in Northern Rhodesia and Nyasaland. The independence of the Congo, formerly Belgian, has spilled its effects into Angola, and the tranquility of Moçambique is being disturbed by the independence of Tanganyika. Whites are divided among themselves over the best method for meeting the events to come, while Africans sense the change of fortunes which now, after many decades, is in the air.

The buffer zone, the wide belt of land where these changes are now taking place, the region which keeps white man's South Africa separated from the immediate impact of the Wind of Change, today consists of a core and two vulnerable flanks. The core is the Federation of Southern Rhodesia, Northern Rhodesia, and Nyasaland, which came about in 1953. The flanks are Angola in the west and Moçambique in the east, giant territories where millions of Africans, who have progressed little over centuries, live under the iron hand of the Portuguese administrators. The youthful Federation (also called the Central African Federation) and the ancient Portuguese possessions are bracing themselves for the greatest crisis in their history. The people, white and non-white, in the various countries have widely varying opinions concerning the most appropriate course of action. Here, a new factor has entered the sphere of political development. More than ever before, Africans within and outside these territories are exerting an influence with which policy-makers have to reckon, whether or not there are official channels through which these Africans can make themselves heard.

| History and Politics

The landlocked Central African Federation is a youthful state. Not only is its present political organization a new one, but the entire history of white settlement covers less than a century. Most of the economic development and political changes of importance have occurred during the last 50 years, with some regions experiencing a growth exceeded only by parts of South Africa. Southern Rhodesia found itself particularly well endowed with resources, including good soils, adequate rainfall in certain areas, and a variety of mineral deposits. Northern Rhodesia's Copperbelt stimulated immigration and exploitation, but in Nyasaland no mineral riches or other wealth attractive to white settlers were located. Hence, the total number of white people in Nyasaland has remained low, while

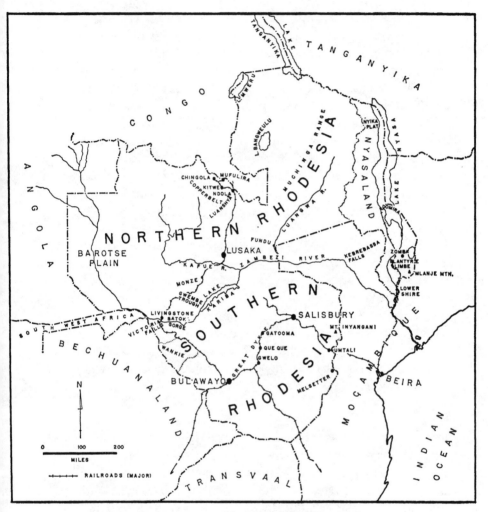

MAP 8 *Federation of Rhodesia and Nyasaland.*

there are a larger number in Northern Rhodesia, concern-
trated in the Copperbelt and along the railroad leading
from this region to the south. Southern Rhodesia attracted
by far the greatest number of white settlers, and they began to arrive early. In the 1890's they were already numerous enough to defeat the African peoples in battle, and by

1914, when they numbered nearly 25,000, some 7,000 of them went to war for Britain in Europe. This figure indicates the youthfulness of the immigrant population. How else could 25,000 people produce an army of 7,000? The Southern Rhodesian settlers, like their contemporaries in the south (from where many had actually come) were hardy people, with much war experience, and they shared the racial prejudice of their southern neighbors. They came to profit from the land and its untapped resources, and they stayed to rule the local African peoples. As in South Africa, the whites in Southern Rhodesia soon began to demand a greater degree of independence from Britain, and as their numbers grew, so did the desire for self-determination. To the north of the Zambezi River, which separates Southern Rhodesia from the two northern territories, the smaller number of whites and the slower rate of development, as well as the nature of white-African relations, caused an altogether different course of events. Here, the Africans had not been defeated by white invaders. There was a process of peaceful penetration, and what violence there was revolved around the elimination of the slave trade. African peoples were left on their lands, their chiefs remained in power, and commercial activities were in the hands of the British South Africa (Charter) Company. No uncontrolled staking ("pegging" in Rhodesian parlance) of claims could take place, and though missionaries, traders, and planters did enter Northern Rhodesia and Nyasaland, they never came in the numbers which characterized white settlement to the south of the Zambezi. Of course, friction was not totally absent in the north. Whites occupied some areas in Nyasaland and began to grow tea, which was found to be well suited to parts of that territory. Tea exports began as early as 1908. With the discovery of the mineral area now known as the Copperbelt, a railroad was begun across Northern Rhodesia, and mines were started; Broken Hill began production in 1902. All this did have its inevitable effects, but African rights were safeguarded and land alienation was minimized. Since the economic development was extremely localized, the majority of Africans were not affected by it. North of the Zambezi, the coming of the white man had a minimum of reverberations.

Political activity, particularly after the formation of the South African Union in 1910, was concentrated in Southern Rhodesia, and there it involved almost exclusively the white minority. World War I, however, had its effects throughout what is today the Central African Federation, and these effects were most severe in the north. In the first place, the north was closest to the East African war theater, and in the warfare waged there, many Africans were involved, mainly as carriers of supplies and ammunition. This caused some dissatisfaction which, coupled with grievances over the slowness of economic development, lack of opportunities, and the treatment of workers on some of the Nyasaland plantations, gave rise to agitation and subversive activity. In Southern Rhodesia, meanwhile, the absence of several thousand whites caused a shortage of skilled labor, and in addition prices rose, an obvious result of the location of the Rhodesias and the war-created difficulties in transporting consumer goods there. There were some strikes and labor unrest of other kinds, but the trouble was confined to the whites. The Africans, who had unsuccessfully revolted at the time of the Jameson Raid, did not at this time attempt an uprising; doubtless many remembered the results of their previous effort.

After the war, significant changes took place in the Rhodesias. In 1919, the Southern Rhodesia Legislative Council raised the matter of responsible government with Britain, which eventually led to a referendum, held among the white population in 1922, to decide whether Southern Rhodesia should assume its own government or should join the Union of South Africa. This was another referendum which greatly affected the people who were not represented in the vote: the Africans. By 8,774 votes to 5,989, Southern Rhodesia voted for self-government. The vote indicated the considerable strength of the Afrikaans element in the territory (about one in five of the white population) and the number of people who even then saw the trend of things in South Africa as the desirable one for their own country. With this vote, Rhodesia became a Self-Governing Colony in 1923. The governing body became the Legislative Assembly, from which a Cabinet was selected. The electorate consisted of the people who could satisfy certain requirements in terms of salary and property ownership. These requirements deprived virtually all Africans of the opportunity to register as voters, and although the country never has had the voters separated according to race alone, the prevalent practices have ensured the whites the dominant voice in the Assembly. With the granting of self-government to Southern Rhodesia, which except for a few minor restrictions was complete, ties between Britain and the territory were loosened. This is what Africans dominated by a local white minority always fear, and in Northern Rhodesia and Nyasaland the close supervision of the affairs of the territories by the British Colonial Office was jealously preserved by the Africans.

The differences which existed in government and race relations north and south of the Zambezi River were intensified after 1923. Already, an effort to merge the territories had failed. Southern Rhodesians did not wish to be encumbered by the costly administration of a large and backward territory where Africans dominated numerically to an even greater degree than they did in the south. Africans in the north retained a great deal of power through their chiefs and councils, and they already had worked their way up into administrative office jobs which, in Southern Rhodesia, were reserved for white men. Their removal from these jobs and the imposition of Southern Rhodesian practices upon this north country would have cost the settlers in the south much tax money and slowed their own progress. In Southern Rhodesia the feeling was, among whites, that if the settlers in Northern Rhodesia and Nyasaland wanted political advancement, they should depend upon themselves to obtain it direct from London. Southern Rhodesia had voted for continued separation from South Africa largely on economic grounds; the whites feared an exodus of African labor across the border to the Witwatersrand with its higher salaries, they feared that Africans in South Africa were getting jobs which in Rhodesia were reserved for whites, and they feared an influx of poor whites, with the problem of destitute whites in South Africa gaining significance while it was still virtually unknown in Rhodesia. Furthermore, there were, in these first decades of the 20th century, still marked differences in the economic progress of South Africa and Southern Rhodesia. South Africa was beginning to protect local industries by tariff walls, while Southern Rhodesia wished to keep taxes and prices of imported consumer goods low. So, although there were people who voted against union with South Africa because they objected to bilingualism and South African racism, economic considerations were paramount. And just as Southern Rhodesia was loathe to join a more advanced partner, so it did not wish to be obstructed in its progress by being joined to backward Northern Rhodesia. Africans in Northern Rhodesia were likewise against any permanent amalgamation. Perhaps they did not understand the exact governmental ramifications of the move, but they did see clearly that, compared to their countrymen in the south, they were privileged people who had a great deal of freedom. They did not wish to see their lands taken away and their local authoritities removed, and to them, the early failure of efforts to unite the Rhodesias was a source of satisfaction.

The whites in Northern Rhodesia, meanwhile, were likewise by no means unanimously in favor of amalgamation. They did not wish to see workers migrate from the Copperbelt, where labor was needed, to Southern Rhodesia. Northern Rhodesian Africans, after all, had not been deprived of the bulk of their land, and thus there was no need for many Africans to flood to mines and cities for a livelihood. Nyasaland plantation owners did not want the competition from the rich agricultural lands of the south for markets which had always been theirs. The Chartered Company, at the same time that the question of self-government was being debated in Southern Rhodesia, began negotiations with the British government concerning the future of the north, over which it had for some decades held sway. While relinquishing its administrative functions to the British government, the Company retained its mineral rights. So, in 1924, Northern Rhodesia came to be a protectorate under the Imperial Government, under the control of the Colonial Office. An Executive Council assisted the Governor in his administration of the territory, while a Legislative Council was also created. These councils were, of course, dominated by appointed senior civil servants, but their very existence established a goal for Africans, a mechanism for government in which they could in time come to participate. Nyasaland, which had been a protectorate since 1891, continued under a similar form of administration. In the years 1922 to 1924, the divergence of social and political development which had begun in the very first years of white penetration in the area of the Zambezi River, jelled in such a manner as to intensify it. South of the Zambezi was white man's land, where Africans were ruled as conquered people and treated much like those in South Africa. North of the great river the whites were simply too few in number to rule in a similar manner, and the influx had been different in kind. White-ruled Southern Rhodesia and missionary-protected Nyasaland were about as far apart on the scale of white-African relations as any territories in Africa ever were. Northern Rhodesia was in the middle but tending toward the side of Nyasaland. Things were later to change in Northern Rhodesia, locally in a violent manner, and there was to be a wave of white immigration. But for three decades, the governmental organization remained unchanged, perpetuating the differences on the two sides of the great Zambezi River.

| *People and Places*

Southern Rhodesia, Northern Rhodesia, and Nyasaland together cover some 486,700 square miles, which is slightly more than the Republic of South Africa without South West Africa and exceeds the combined area of Texas, California, and New York state. The combined population is between 8 and 9 million, of whom over 300,000 are white. This is only about half the South African population, and the white sector is less than one-tenth as large as that of the Republic. In the whole of the Federation, there are less than 40,000 people who are neither African nor white, and thus there are no sizable Colored or Asiatic minority groups. Of the three territories, Northern Rhodesia is by far the largest. It is about twice as large as Southern Rhodesia, while Nyasaland's 46,000 square miles (about the size of Tennessee) includes over 9,000 square miles of Lake Nyasa. Nyasaland's land area, in fact, would fit eight times into that of Northern Rhodesia. Yet in total population, Northern Rhodesia has the smallest share, and small Nyasaland vies with large Southern Rhodesia for the lead. Some of the troubles in the young Federation become immediately more understandable when it is realized that there are nearly 3 million people in Nyasaland (an average of about 80 people for every square mile of land) where resources are fewest and progress slowest. Southern Rhodesia also has somewhere near 3 million people (19 people per square mile), and here economic development has been most rapid. Northern Rhodesia has well under 2½ million people, or about 8 persons per square mile. These averages are, of course, not very meaningful if applied to small areas in each territory, but it is obvious that pressure on the land in Nyasaland is considerable. The one territory among the three that is virtually entirely agricultural is also the most densely populated. This means that the high population of Nyasaland is not made up of large urban agglomerations; in fact, in this territory the largest town, Blantyre-Limbe, has less than 80,000 people!

On the other hand, Southern Rhodesia, which has a great deal of good agricultural land and no dense population, does have big cities, such as Salisbury with an estimated population in its metropolitan area of 220,000 and Bulawayo with just under 200,000. In sparsely populated Northern Rhodesia, the only sizable town that does not lie along the Copperbelt is the capital, Lusaka (80,000).

The distribution pattern of the white population taken alone is very different. Of the 300,000 whites in the entire Federation, about 72 per cent live in Southern Rhodesia, 25 per cent in Northern Rhodesia, and only 3 per cent in Nyasaland. Whites are still arriving at the rate of about 10,000 per year in Southern Rhodesia, while the immigration rate into Northern Rhodesia is very small and into Nyasaland negligible. The large number of whites in Southern Rhodesia must be attributed to the opportunities the land provides and the policies of the local government, which has deliberately encouraged immigration with the same object as South Africa. Elsewhere, there are only civil-service jobs and limited industrial and mining development, so that of the three Federation territories, Southern Rhodesia, with its available land, good soils, good climate, and considerable resource base, has been by far the most attractive to immigrants. The white population is highly urbanized. Salisbury and Bulawayo together form the place of residence for nearly half of Southern Rhodesia's 216,000 whites!

The three territories all lie within the tropics, but they stretch mainly over the African plateau, so that they are relatively cool, like South Africa's highveld, because of elevation. In fact, the bulk of the entire Federation lies above 3,000 feet, and most towns are at approximately 4,000 feet above sea level. Salisbury, for instance, is located at over 4,800 feet, only a few feet lower than Johannesburg. Bulawayo is at 4,400 feet. The lowest areas are the hot and humid valleys of southern Nyasaland and the valley of the Zambezi River. Livingstone, once the capital of Northern Rhodesia and situated just above the Vic-

toria Falls immediately to the north of the river, is at
3,000 feet. In the south, the land falls toward the valley of
the Limpopo River, which forms the boundary between
Southern Rhodesia and South Africa. The land also falls
westward into the basin of the northern Kalahari and
Bechuanaland, but the deep river valleys remain the lowest
points. The surface of Lake Nyasa is at 1,550 feet, and so
some of the land adjacent to the lake is low lying. In the
extreme south of Nyasaland, the Lower Shire is in reality a
part of the Moçambique coastal plain, and this lowland
stands apart from most of the remainder of the Federation.

High regions also are found in the east. Mount Mlanje
in Nyasaland reaches 10,000 feet as it rises spectacularly
from the lowland plain. This is the highest mountain mass
in the country. Most of the high areas of the east form part
of the Great Escarpment, which here as in South Africa
forms the sharply defined edge of the plateau. In fact,
Southern Rhodesia's eastern boundary virtually corre-
sponds to this escarpment, and the land here falls rapidly
from over 5,000 feet in many places to the coastal plain
below. Southern Rhodesia's highest point is magnificent
Mount Inyangani, which exceeds 8,500 feet. Northern
Rhodesia possesses fewest topographic extremes. Apart
from its backbone, the Muchinga Range which runs from
northeast to southwest in the eastern region, the territory
is rather even and undiversified. Scenically, the largest of
the three territories is least varied.

Like South Africa, the Federation possesses a core plateau area which, not surprisingly, resembles the South African highveld. This is a ridge, also running northeast to
southwest, across Southern Rhodesia, and upon it Salisbury, Gwelo, Bulawayo, and other towns are located. This
ridge is not very prominent, being rather a gentle but per-
sistent rise, but it is the region where most of the people of
Southern Rhodesia live and where most of the industrial
development has taken place. This is also the area with
perhaps the best climatic conditions in all the Federation.

Salisbury is warm in October and November, when the temperature averages about 70°F., but it is cool in July, with an average of 57°. In summer, temperatures sometimes reach the 90's, and in winter, frosts occur at night, though they are seldom severe. The climate is warmer, but with the large amount of sunlight and considerable extremes is very reminiscent of Johannesburg's. The total annual rainfall is about 33 inches, most of which falls during a pronounced wet season lasting from November to March. This sort of climate is attractive to white men who come from temperate climes elsewhere.

Because of the size of the Federation, a good deal of climatic variation between various parts is to be expected. The rainy season generally lasts from November to March or April. In some areas, particularly in the higher mountain regions of Nyasaland, the rainy season is longer and the total rainfall higher, while in others, there is so little rain that a rainy season cannot really be spoken of. At Mount Mlanje in Nyasaland, about 100 inches of rain falls, but in the southern part of Southern Rhodesia, along the low, dry border with Bechuanaland, there is as little as 12 inches. It may be said that the areas between 4,000 and 5,000 feet—which is where most whites have settled—generally receive between 25 and 50 inches of precipitation annually. Rainfall in the Federation does not come in slow, drizzly, dreary days. It generally falls in showers which are followed by a clearing of the skies. Along Lake Nyasa, onshore winds bring much rain to nearby areas.

In the north, the rainfall is higher, and in many parts of Northern Rhodesia there is so much that the soils are leached and can support only a shifting kind of cultivation, which means that a small total population is supported. Much of the northern part of Northern Rhodesia is swampy and hot, with bilharzia and tsetse fly still rampant, cattle-ranching impossible, and malaria a constant danger to man. Here, also, it is hotter, and there is less variability in the weather than there is in Southern Rhodesia. The tropics become least bearable in terms of climate where the

monotony of high temperatures and humidity become enervating, and this is the case in parts of Northern Rhodesia.
Particularly the low regions around Lake Bangweulu near
the Congo border, the Luangwa River Valley, and, of
course, the Zambezi River Valley are very hot almost
throughout the year, averaging over 70°F. The shores of
Lake Nyasa and the Lower Shire of Nyasaland belong in
this category, though here the variability is somewhat
greater. While most of Southern Rhodesia is cool with
moderate rainfall, much of Northern Rhodesia and Nyasaland is hot with excessive rain. The elevation of the Copperbelt (mainly over 4,000 feet) and the Muchinga Range
in Northern Rhodesia and of the Nyika Plateau in Nyasaland ameliorate conditions somewhat, but where the land
falls, climate deteriorates. During the building of the famous Kariba Dam on the Zambezi River, temperature in
the shade was recorded to have risen over 130°F. The
Zambezi rises in Angola but crosses a large part of western Northern Rhodesia, and this is one of the hottest parts
of the region. Climatically, the effect of the wide and deep
Zambezi Valley is to separate the cooler sections of the
Southern Rhodesian plateau from the more desirable parts
of Northern Rhodesia, this being coupled with a reduction
of annual precipitation.

The scenery of Southern Rhodesia is not unlike that of
the northern highveld of South Africa. This is mainly
open savanna country, aptly described as "parkland" savanna because of the spacing between the trees, which gives
the country more of a woodland and less of a bushland appearance. Between the trees, which grow to an average of
thirty feet, and interspersed here and there with the characteristic baobab, is tall veld grass which dries out or
"burns" in winter to be replaced in the early part of the
rainy season with new, bright green growth that unfortunately remains green for all too short a time. Mostly, the
aspect of the country is one of grayness, with the sun beating down mercilessly in an apparent effort to wither all
beneath its glare. This grayness disappears in the lower re-

gions of the Zambezi, Luangwa, Shire, and other river valleys. Here, the Mopani tree is most common, and it is a
favorite with elephants. Baobab trees are also most prevalent here, while the highveld grasses are absent and the
ground much more bare. This is lowveld country, with
palms and ferns, resembling the lower reaches of the Limpopo and Crocodile Rivers further south. The densest part
part of this flora is confined to the areas immediately adjacent to the water courses, where also the animals are
concentrated. These low, hot, forested regions, of which
the Federation has many, provide a reminder that this is
interior Africa, perhaps at its most lush. Again, the highlands have their own individual character. The Nyika Plateau, Mlanje Mountain, and Inyangani all possess ever-
green vegetation types: here, indigenous conifers grow.
Some have compared these forests to those of the Cape in
South Africa, and with the magnificence of the mountains,
the comparison is justified.

The Federation does not have a Cape and Natal coast,
a ridge and valley topography such as the Cape Ranges, and
no Drakensberg. However, although perhaps less diversified than the Republic to the south, dominated more by
mile upon mile of savanna and tall grass and subject to less
climatic variability, the Rhodesias and Nyasaland have
some incomparable scenic assets. The magnificent teak forests of the western part of Northern Rhodesia are unequaled, and in the swamplands of the area around Lake
Bangweulu are vast papyrus areas. The shores of Lake
Nyasa and the sheer slopes in the immediate hinterland
leave an unforgettable impression. Vistas from the edge of
the Great Escarpment across the coastal plain are among
the most spectacular in the world. The Wankie Game Reserve, covering 5,000 square miles in the northwestern part
of Southern Rhodesia, is one of the finest of its kind in all
Africa—according to many, superior even to the Kruger
National Park. But most magnificent of all are the great
rivers, flowing quietly along like the Kafue at Blue Lagoon,
rushing ominously forward as does the Luangwa near

Fundu, or plunging spectacularly over huge rocks, like the
Zambezi at Victoria Falls. The Federation is the land of
great rivers. There is nothing in all Africa to compare to
the great Zambezi River, its changes of character, its un-
predictability, and its eternal magnificence. To the country,
these rivers are the arteries of life. Animals trek for uncounted miles to drink along the banks, and vegetation
thrives there, harboring a wealth of fauna. Crocodiles and
hippos lurk in the slow-moving water which each year rises
and falls with the change of the seasons. Man has dammed
the greatest of all—the Zambezi—and has put the rushing
water to his use. He too has come to recognize his dependence upon these great sources of surface water. There
are other places where dams will be built. The Kafue at
Meshi Teshi and the Shire between Nyasa and the Zambezi
—there are dozens of opportunities. Meanwhile, the rivers flow along, their spectacular scenic beauty and power
representing a heritage of millions of years.

The deep, hot, densely forested valley of the Zambezi
River divides Southern Rhodesia from the more tropical,
distinctly different north. In many countries, a river has
had a unifying influence, has become an artery, a focal
point, a transportation route. Rhodesians would like to see
the Zambezi play a similar role. For decades, the river has
divided north and south. It has never been easy to cross,
and the bridge at the Falls was not opened till 1905. Over
much of its course, the great river flows through deep
gorges and remote lowlands; only above the Falls does it
offer a rather shallow and wide valley. Above the Falls,
however, it does not have its dividing function, because the
Falls lie very close to the western extremity of the North-
ern-Southern Rhodesia boundary. There is nothing to equal
the magnitude and impressiveness of Victoria Falls. They
defy description, and Livingstone, who in 1855 was probably the first white man to reach them, exclaimed that
"scenes so lovely must have been gazed upon by angels in
their flight." Emitting a roar that can be heard for miles,
producing a spray visible 20 miles away, adorned by rain-

bows, the mighty Zambezi plunges over a ledge that is well over a mile wide into a narrow gorge more than three hundred feet below. The greatest fall is 355 feet, or well over twice the height of Niagara Falls. In April and May, the river reaches its maximum flow, and at that time between 75 and 120 million gallons per minute cross the Falls' ledge. The mist is then so dense that the Falls are obscured except when a breeze temporarily displaces the mantle of spray. At low water the Falls lose somewhat in spectacle, but the advantage of clearer views accompanies this period of relative quiescence. The Falls have been retreating along a series of lines of weakness in the basalt rock, and the future position of the Falls and its receptacle gorge can be determined. The rock is resistant, however, and retreat is not rapid considering the tremendous power of the river. Below the present Falls are a series of gorges marking the previous locations of the main Falls, and these gorges, because of the geologic peculiarities of the rocks, lie in a zigzag pattern, the water rushing at great speed around the narrow hairpin bends. These gorges themselves are spectacular achievements of nature. In addition, there is much wildlife in the immediate vicinity of the Falls, and other features of great interest may be seen, such as peculiar types of vegetation, sites containing stone implements left by prehistoric man, and magnificent walks and drives in the surrounding area. Notwithstanding this variety of attractions, the Victoria Falls area has not become excessively commercialized. Though it is the major tourist attraction of the Federation and annually draws thousands of visitors, much of its beauty is as wild, unspoilt, and breathtaking as it was when Livingstone first saw it.

| Damming the Zambezi

At Victoria Falls, the Upper Zambezi ends, and with it the open, shallow, accessible valley. Below the Falls is the Batoka Gorge, the first of a series of gorges through which the Middle Zambezi rushes on its way to the Kebrabassa

Rapids, which lie 360 miles upstream from its delta. At Kebrabassa (well inside Moçambique) the Middle Zambezi terminates, and the river assumes a winding, slow-moving path across the flat coastal plain. Thus the entire course below Victoria Falls, the course of gorges, lies between Southern and Northern Rhodesia, and here the Zambezi has served to separate the territories historically as well as physically. Hot, deep, inaccessible, unexplored and repelling, the larger part of the Middle Zambezi Valley had defied penetration. Here, however, the river afforded opportunities for dam development. Rhodesians were long aware that the building of a large dam would have the effect of flooding a large part of this undesirable course of gorges, could supply electricity to north and south alike, would bridge the river and form a belt of contact between the two territories. In short, the river could become a unifying factor in the region rather than the dividing one it had so long been. But the damming of the Zambezi was no small task, and for a country still developing its economy, the building of such a great project was for long impossible. In addition, there were other, possibly less costly, sites to be considered. The Kafue affords several likely places, and other rivers deserved attention. Why the unparalleled scheme on the Zambezi, with all its risks, financial and technical?

In due course, the building of a major power project became a matter of urgent necessity. Africa south of the Sahara is notoriously poor in mineral fuel resources, and though the Rhodesias possess extensive deposits of coal, these are very localized. Hence the coal must be hauled over long distances to power plants by the railways which are already overtaxed. It was feared that the railroads would be incapable of supplying industries and mines with coal unless astronomical expenditure took place in the purchase of new equipment. In addition, the transmission problem, long a vital deterrent to the development of distant hydroelectric sites, was being overcome, electricity now being transmitted over hundreds of miles. Finally, the economy of the Rhodesias continued its progress, so that by 1950 it seemed that the money necessary for the building of a major hydroelectric power plant could be raised.

The actual selection of the site for this great project, which engineers foresaw as one of the great feats of history and politicians heralded as the beginning of a new era for the country, caused some friction. Northern Rhodesians favored the site of the Kafue River at Meshi Teshi, where there is a 1,900-foot drop over a distance of 22 miles. However, more was known about the regimen of the Zambezi River, its silt content, extreme volumes, and other characteristics, while the Kafue was relatively unknown. In addition, a Kafue project would have drowned some excellent agricultural land, while several of the gorges of the Middle Zambezi were sparsely inhabited, virtually undeveloped, and desirable for dam sites from other points of view. The gorge selected was that of Kariba, also known as the Gwembe Trough. The project planned was the greatest ever devised by man. A dam wall nearly 2,000 feet in length and over 400 feet high was to be built across the mighty Zambezi. It was to make a lake some 175 miles long and up to 20 miles wide. The capacity of this dam was to be 130 million acre feet, or about 4½ times that of Hoover Dam. The surface area of such a dam was estimated to be about 2,000 square miles. A deep gorge would be replaced by a wide lake, which brought visions of a fishing industry, thriving agriculture along the banks, and a new transportation route in the heart of the Federation.

Almost as soon as the Kariba site and plan were approved by the Federal Parliament, work was begun. Ground was broken in June, 1955, and today power is flowing from the completed dam. Africans warned that the Zambezi would not succumb without doing battle with the dam-builders, and many predicted that this was where the white man would fail. Economists still feared the effect of the project effort on the economy, in that certain materials such as cement and building materials would be in short supply while Kariba was being constructed, and the transportation system would be temporarily overloaded. The total cost of the first stage of the scheme was about $224 million, and this rose to $320 million before the dam was completed. Of the initial sum required, the International Bank loaned over one-third, and the Copperbelt companies, who were to profit considerably from the new supply of power, provided a quarter. Other sources of funds were the Colonial Development Corporation, the British South Africa Company, various banks, and the Federal government, which promised to provide the difference between the loans acquired and the amount needed for completion of the first stage.

The Zambezi did not submit to the engineers without, it almost seemed, proving those correct who had predicted disaster. Soon after the project had begun to progress significantly, the river rose to its highest flood level in recorded history, invaded the protecting walls of the so-called coffer dam (circular walls within which the actual dam was being constructed), destroyed equipment, washed away whole sections that had been completed, and caused destruction at the site. The work was set back for months.

By this time, Kariba had begun to look like a new town. The Zambezi had been bridged by a temporary structure, roads were laid through the previously isolated country, airstrips were constructed, homes and apartments were built for the 1,000 whites and 6,000 Africans working there, hospitals were set up, and the whole area, desolate and deserted all the years previous, came to noisy, bustling life. The problems to be faced were many and varied. Not only was there an urgent need to combat malaria and other diseases in the area where whites fresh from Europe were at work alongside native Rhodesians, but here, 240 miles from Salisbury and 120 miles from Lusaka, there was no agriculture, and food had to be brought in. Temperatures in the hot season rose to intolerable levels. Elsewhere, meanwhile, the effects of the dam wall, now under construction, had to be anticipated. The valley to be inundated for over 175 miles was occupied, if sparsely, by an African people called the BaTonka as well as a wealth of fauna. Some 50,000 people had to be moved to other areas while the water was still low; once it began to rise, there would be no time. The resettlement alone was a major project. In addition, it was imperative that at least some of the vegetation of the lower gorge be cut and burned, and an investigation showed that a clearing of vegetation on the Southern Rhodesia side to a depth of sixty feet would cost between $5 and $6 million. At the dam site itself, the problem was not only the building of the dam wall but also the blasting, cutting, and completion of the turbine tunnels and chambers through which the water was to flow when the lake filled up. Heavy machinery and equipment had to be moved along roads not fit for the purpose, and the wet season, while delaying activities, could not be permitted to halt this transport—or that of food, drinking water, and innumerable other items. While the turbines were being installed, transmission lines must be laid along nearly 1,000 miles, and for this, nearly four thousand pylons had to be built. This involved the cutting of a wide swath through the wild country, the transporting there of the equipment for setting up the pylons as well as the distribution of the sections of the pylons themselves. The initial part of the transmission system was to connect Kariba with Kitwe in the north on the Copperbelt and with Salisbury, Norton, Umniati, and Bulawayo along the plateau ridge of Southern Rhodesia. These projects all had to be carried out simultaneously, and there was no time to be lost.

Just over 2½ years after the first concrete was poured in the foundations for the dam site, the great turbines delivered their first units of power. In spite of all obstacles, the main part of the work was finished at the end of 1958. Lake Kariba began to fill up, watched in awe by scientists, engineers, and laymen alike. While all the people had been moved out of the Kariba gorge, there was of course no way in which an exodus could have been organized to drive out the animals. Rhodesians watched with concern how animals in the gorge made for the islands, which then disappeared

as the water engulfed them. Volunteers came from everywhere—game rangers, students, holiday-makes—all intent
upon helping the trapped victims of Lake Kariba. Whites
and Africans worked together day and night in driving the
larger animals back into the water, while launches herded
them to safer land. Small animals were trapped and caught,
bound, and transported to the mainland, there to be released. Even dangerous snakes, wildly kicking zebra colts,
hares, lizards, and tortoises—as many as could be saved—
were rescued. "Operation Noah," as this rescue effort has
come to be known, proved that Rhodesians appreciate their
wildlife heritage and will go to almost any effort to preserve it. "Operation Noah" gained possibly more fame
than Kariba itself, as newspapers, magazines, and newsreel
and television cameras captured the drama of man helping
the innocent victims of his works.

It is not yet possible to assess the total effect of Kariba
upon the Rhodesias and Nyasaland. The lake is now filled,
and power is flowing to mines and industries. The cost of
the electricity will drop constantly as loans are repaid. The
railroads are burdened with about a million tons of coal
less than they would have been today without Kariba. Most
of all, however, putting the Zambezi to work on such a
monumental scale shows what can be done in Africa. The
Kariba Dam was planned as long ago as 1925. Now that
it has been a success, other sites are already under investigation, and the Kafue and Shire may in time supplement the
Zambezi in supplying hydroelectric power. Great schemes
have been planned elsewhere in Africa, such as in Ghana on
the Volta River and in the Congo at Inga. Kariba is the
first project of its magnitude completed in this continent,
which has more hydroelectric power potential than any
other land-mass on earth. It was built not for reasons of
national prestige but because it was urgently needed in the
economic development of the Federation. Urgent need
causes action where it may have been long delayed for lack
of impetus. Many of Africa's other dam sites are far from
present markets for power and would cost so much as to be

out of the question for the countries involved. When, however, the time comes for these other sites to be developed, Kariba and the lessons learned there will be a most valuable asset. In the meantime, Kariba will form something of a monument to enterprise and far-sightedness in the Rhodesias. The Zambezi has been bridged in yet another place, for a wide roadway runs along the top of the wall of the dam. The lake is the largest man-made body of water in the world. Whether Kariba will have a unifying influence upon the two Rhodesias and Nyasaland remains to be seen; whether its effects will be as predicted cannot yet be ascertained. Significantly, perhaps, in a country that ranks among the world's leading copper producers the transmission conductors are made of steel-cored aluminum, a reminder that the refining of the copper, which could give employment to many, is done in another country. The electricity, moreover, is put to different uses north and south: in Northern Rhodesia it serves the mines, in Southern Rhodesia the industries.

Land and Resources

The resources of the Rhodesias and Nyasaland are many and varied, but they are unevenly distributed. One of the main arguments in favor of federation has always been that amalgamation would aid the poorly endowed areas while better balancing the economy of the more prosperous areas. Of the three territories, Southern Rhodesia has most of the wealth, which is perhaps an adequate explanation for the large number of white settlers there. But even before the coming of the white man, mining took place there. Ancient, now deserted workings of gold and copper as well as iron have been found, and possibly millions of ounces of gold were taken from what is today Southern Rhodesia before the modern mining phase began. It was gold which attracted a number of the early settlers to Southern Rhodesia, and in fact gold continued to be the mineral producing the highest revenue until as late as 1929, when for the first time other minerals, discovered over the years, exceeded gold in production returns. The mining of bulk base metals and other minerals was retarded until the railroads reached the regions where the ores are located. Coal, for instance, was discovered at Wankie (near the Game Reserve) as early as 1894, but it was only ten years later that production on a major scale began. In addition to gold and coal, deposits of asbestos, chromium, lithium, tin, copper, iron, and nickel have been located. Asbestos frequently exceeds gold in value of output. In high-quality chrysotile asbestos, Southern Rhodesia is the world's largest producer, and in total production of all kinds of asbestos it ranks third. In world gold production the territory is in seventh place. In chromium, also, the area is well endowed, and production is limited only by the capacity of the railroads to haul the bulk material. Chromium is in rising demand on world markets, and Southern Rhodesia again ranks among the top producers. Lithium, likewise, has been in increased demand during recent years.

The existence of coal deposits and iron ore in Southern Rhodesia has been of great local importance. Coal is mined in quantity only at Wankie, and there the quality is good enough for coking. In addition, the iron-ore deposits of Southern Rhodesia may rank among the world's largest, so that the present iron and steel industry of the Federation, which is confined to Southern Rhodesia, is likely to grow. This places the country in an enviable position (with South Africa) in subsaharan Africa. Particularly with the decreased demand for coal by thermal plants producing electricity, and with the present investigation of high-grade iron-ore deposits located near the railroad to Lourenço Marques, there is every reason to predict a faster growth for this important industry. There are abundant limestone deposits in several parts of the territory, so that the necessary ingredients are there, virtually without limitation. The main steel plant is situated at Que Que (about midway between Salisbury and Bulawayo). In the past few years, production has more than doubled, from 75,000 tons in

1958 to about 150,000 tons in 1960. Should the Wankie Colliery be unable to furnish the required supply, further coal deposits located in the east may be utilized.

In addition to these major mineral deposits, Southern Rhodesia has a large number of relatively unimportant mineral deposits which, of course, might attain significance in the future. Lead, platinum, silver, tungsten, and a host of other minerals have been discovered on the plateau. This astounding variety is largely the result of a geologically unique phenomenon: the so-called Great Dyke. Minerals often form as a result of the invasion of new, liquid rock into pre-existing rock layers, the contact zone being one of chemical activity. Such occurrences are usually limited in extent. In Southern Rhodesia, however, there is one 320 miles in length and averaging from 3 to 6 miles in width. It runs from northeast to southwest along the plateau, forming a gentle rise so that in places it is quite conspicuous. Whether it is actually technically a dyke remains open to debate, but there is no question concerning its effect. All along the lengthy ridge, mineral deposits occur, and the variety is considerable. This is one reason that white settlement in the territory is concentrated along the similarly trending backbone of the plateau. The Great Dyke is Southern Rhodesia's greatest blessing, and its influence on the course of events has been considerable.

Northern Rhodesia has always had a far greater reputation in terms of mineral production than its southern neighbor, even though the mineral diversification here is much less. This territory's fame rests, of course, upon the Copperbelt, which is the southwestern half of the mineralized complex known in the Congo as Katanga. The economy has always been dominated by copper since ore was first produced in 1902—several years before the railway from the south reached it. The degree of domination of copper production can be assessed by comparing the total value of all mineral production in Southern Rhodesia for 1958, $70 million, to the value of copper alone in Northern Rhodesia, $224 million. Cobalt, lead, and zinc are mined in the Copperbelt where they are found associated with the copper. There are some minor minerals also, but none are of any significance in the total production. The copper industry of Northern Rhodesia is the most important single enterprise of the new Federation, contributing as much as 20 per cent of the state's fiscal revenue. Production only began to increase significantly after the important political changes of 1923, when Northern Rhodesia's status as a protectorate was established and the British South Africa Company retained possession of mineral rights. Once unencumbered by administrative problems, the Company began to encourage exploitation in the Copperbelt by granting prospecting rights. Important discoveries were then made, and mines were founded at Luanshya, Kitwe, Mufulira, and Chingola, while electrolytic refineries were also built. The major developments in the Copperbelt took place after 1924, and this fact is of importance in the political evolution of the entire Rhodesia—Nyasaland region.

Nyasaland is least fortunate in terms of mineral resources. There are a number of minor deposits, but none is of sufficiently high grade or extent to permit profitable exploitation. Yet Nyasaland, with its high total population, would benefit greatly from a mining industry. Several factors have prevented the establishment of such an industry, and the apparent absence of workable deposits is only one. Nyasaland's location, transportation costs, and the high price of electricity would make it extremely difficult for any mining company to export a product that could compete on the world market.

In terms of mineral resources, therefore, Northern Rhodesia is by far the best endowed of the three territories, and the revenue derived from the copper, though subject to fluctuating world prices, is always considerable. Apart from copper production, however, Northern Rhodesia does not have any sizable source of income, and in both industrial development and agriculture, Southern Rhodesia is more advanced than either Northern Rhodesia or Nyasaland. Southern Rhodesia has the better soils,

more desirable rainfall conditions, and in the growing
cities markets have developed to stimulate agriculture.
Much research into agriculture is carried on here, where
the crop variety is considerable. Irrigation is practiced
where rainfall is low and variable, and some very large
schemes are planned for the future. Throughout the Federation, corn is the staple food crop, and a surplus is produced in Southern Rhodesia and Nyasaland. In Nyasaland
particularly, the storage of excess production to be used in
times of crop failure is now being practiced, and there are
also large silos for this purpose at Salisbury in Southern
Rhodesia and at Monze, Lusaka, and Kitwe in Northern
Rhodesia.

There is, however, much difference between agriculture
carried by whites and by Africans. The latter are still
largely, though not entirely, engaged in subsistence farming, and this is particularly true in Northern Rhodesia. In
Nyasaland, where the fertile soils do not extend over a
large area, population density on the agricultural land
averages nearly 330 people per square mile (less than 2
acres per individual) and varies between 100 and 800. Although some cash crops are grown by Africans even in this
densely populated country, the bulk of the land simply has
to be put to work in producing food for the people. In
neither Nyasaland nor Northern Rhodesia is much of
the land in the hands of whites, so that cash crops, although
they are produced, are not grown in such quantities as to
attain world significance. Nevertheless, even in crowded
Nyasaland some African farmers make a little profit out
of tobacco and cotton, and in Northern Rhodesia some
African farmers grow groundnuts (peanuts) in rotation
with corn and make some profit on this crop also. In Southern Rhodesia, where there has been much land alienation
by the whites, the production of cash crops is virtually entirely in the hands of white farmers, and the Africans can
attempt only to grow sufficient staple crops for their own
consumption.

Much of Southern Rhodesia's land belongs to white

farmers because a great deal of it is of good quality. The suitability of soil and climate for the cultivation of tobacco was recognized in this territory as early as the 1890's, when experimental plantings met with success. Today, cultivation of tobacco is the most important single industry in Southern Rhodesia, and the revenue derived from to-bacco is second in the entire Federation only to that of copper. The value of tobacco exports in an average year is over three times that of asbestos and no less than four times that of gold. Production in recent years has increased tremendously. In Southern Rhodesia, the 1949 production total was 90 million pounds, and that of just ten years later was nearly double this amount. Salisbury, with its three large auction floors, is the largest marketing center for tobacco in the world. The tobacco is exported largely to Britain and Australia but also to nearly 40 other coun-tries. It is estimated that only about a quarter of the suit-able land is under the crop and that expansion will continue, dependent upon world demand.

A large number of other crops are grown on Southern Rhodesia's white farms, though as a cash crop tobacco dominates all. The staple is of course corn, and it covers a large acreage. In addition, barley, potatoes, groundnuts, cotton, and wheat are cultivated, and although none of these attains any great importance, they have all proved their capacity for growth in this region. In fact, there are few crops that have not at one time or another been grown in Southern Rhodesia, and an efficient experimental station at Salisbury has analyzed each with reference to the envi-ronmental conditions of the country. Sugar has been intro-duced, and within the next few years the Federation should become self-sufficient in this respect, as irrigation schemes are completed. The white farms also produce tea, for which Nyasaland is much more famous than is Southern Rhodesia, though in recent years tea production in the latter country has expanded rapidly. Within the Federa-tion, Southern Rhodesia has the best of almost everything. The increase in returns from cash crops have been possible, naturally, because so much land is in the hands of whites. It remains difficult to encourage Africans to grow crops other than subsistence crops, and in Southern Rhodesia the view that "they did not do anything with the land when they did possess it" is an oft-quoted justification for the land alienation that has taken place there. Southern Rhodesian whites also point to Northern Rhodesia, where Africans do have most of the land and where the subsistence habit has been hard to break.

Northern Rhodesia has a very much smaller white population than the south, and the great majority of the whites are involved in mining, administration, and other nonagricultural activities. Whereas Southern Rhodesia's white farms are scattered all over the country, the white farms of the north are virtually confined to the strip along the railroad connecting the Copperbelt with the south. Corn and tobacco are grown, and in the vicinity of Lusaka are some vegetable and fruit farms. In the total picture of the Federation, however, Northern Rhodesia's agricultural production is insignificant. Nyasaland, on the other hand, does have important cash crop production, even though the land in the hands of white farmers is not extensive and the remainder, in African hands, is overcrowded. The whites grow tea on large estates totaling about 30,000 acres. The crop was introduced here as early as 1878, but the industry met with failure for some time. Then, in 1904, the first export of only a few pounds was made, and today an export total of 25 million pounds would be normal. In value, tea ranks just behind gold in the exports of the Federation, and in Nyasaland itself, it ranks second, behind tobacco. Tobacco is one crop which Nyasaland Africans are producing much of, and the value of Nyasaland's tobacco crop exceeds that of tea. In 1958, there were 73,000 registered African growers of tobacco.

Like the Transvaal and the Orange Free State, much of Southern Rhodesia is good cattle country. In fact, of the 5 million cattle in the Federation today, over 70 per cent are in the former colony. While Southern Rhodesia is largely free from the dreaded tsetse fly, this pest still prevents the keeping of cattle over most of Nyasaland and Northern Rhodesia. This results in a concentration of those cattle that are kept in Northern Rhodesia to a few localities, including the western Barotse plain and the eastern highlands. Nyasaland does not have the space for a large livestock population, and those cattle that are kept are found mainly in the central and northern part of the territory. Thus, Southern Rhodesia once again dominates the scene. Here, however, the best grazing lands lie on the white lands, although many more cattle are owned by Africans than by whites. Southern Rhodesia's Africans have come increasingly to feel the pressure of population on limited land. When much land is alienated, some arrangement must be made to accommodate the population which cannot any longer practice its shifting and admittedly destructive subsistence agricultural and pastoral activities. The white man not only took much of the land, but he terminated tribal warfare, improved health conditions, and checked famines. As a consequence, the African population in Southern Rhodesia grew rapidly, and it soon began to exert pressure upon the bounds imposed upon its living space. The cattle population of Southern Rhodesia have meanwhile also increased because of white-imposed practices of dipping and immunization, and the Africans came to need more land than they required when first contact was made with the invading whites. The African had never taken much care of his land, and it cannot be denied that he made no real effort to rise beyond the subsistence level. He unwillingly adopted some conservation practices, but undoubtedly when he began to demand more land, it was first a matter of necessity, all other considerations coming later. In the eyes of the white man, who possessed most of the land by claim or concession, the African had not used his own land in a manner that justified the relinquishing of white acreage.

In 1930, a Land Apportionment Act froze the land dispute in such a way that 48 per cent of all land went to the whites (today numbering 220,000 and then much less),

42 per cent was allotted to the African population (today about 3 million), and the remaining 10 per cent went to game reserves, forest preserves, and other miscellaneous uses. This Act, instead of providing more land to Africans, actually caused nearly 120,000 Africans to move from the white lands between 1936 and 1960, further overcrowding the African areas. There has, of course, never been anything like this in Northern Rhodesia and Nyasaland, and the Africans there feared federation because they did not wish to be dominated by those whites who had put the Land Apportionment Act into effect. While the effects of the Act were accumulating and the African lands deteriorated, plans were devised to aid the African in making the best possible use of his areas. Among the salient conditions prevalent in the African territories was the ever-increasing fragmentation of the arable land and consequent overgrazing and overexploiting. Great numbers of the able-bodied men were away at work on the mines and in the cities, and they would retain their small parcel of the reserve. Upon marrying, they would leave their families to cultivate these plots, which were then divided among children as the families grew, eventually reducing the units in size to far below practical proportions. To combat this and other problems, the Native Land Husbandry Act was promulgated in 1951. This Act, by registering land rights in the names of individual land-holders and thus encouraging pride in ownership, aimed at providing to the land-owners a new direction in soil conservation, intensive farming methods, and a general incentive to better practices. The Act contained powers for the enforcement of its stipulations. But to Africans, who have for so long been accustomed to communal land ownership, these new plans did not seem to offer any new hope. Neither did those measures designed to improve cattle strains. Cattle would have to be killed in order to give the best animals more living space and pasture, but such destocking is utterly opposed to African practices and beliefs. Cattle are counted in numbers and as such are a measure of wealth; they also

322 | AFRICA SOUTH

possess a religious significance here as they do among the
Zulu in South Africa. Destocking was not a practice in
which the cooperation of Africans could be expected. To
most Africans, the answer to their problems remains the
opening of white lands to all races. They already occupy, as
squatters, about 2 million acres of white land, and they desire more.

The Prime Minister, Sir Edgar Whitehead, in 1961 attempted to forestall the friction which was likely to erupt
into violent action by Africans in support of their demand
for more land. The Land Apportionment Act was amended
so that the 2 million acres already occupied by African
squatters became African land, while an additional 5 million acres was to be thrown open to all races at a later
stage. Sir Edgar was acting in response to the report of a
commission which in 1960, after two years of study of the
land problem in Southern Rhodesia, had stated that the
Land Apportionment Act should be repealed in stages and
that 90 per cent of the land should be opened to all races.
This would virtually eliminate the white areas, and to many
Southern Rhodesian whites this program is unthinkable.
But so was a similar program in Kenya, where the invasion
by Africans of the White Highlands seemed similarly beyond contemplation, and yet it came to pass. The important point in Southern Rhodesia is not that the repeal of
the Land Appointionment Act is opposed by whites but
that the pace of repeal will be too slow to satisfy the Africans. Having once established the idea of land-holdings, the
Southern Rhodesians now find that there is not enough African land to allot to all who are eligible, and about 111,-
000 African families are without what in terms of the Land
Husbandry Act they are entitled to. All this happens while
millions of acres of good land, owned by whites, lie unused
year after year. As long as this is the case, it is not likely
that Africans will be disposed to listen to admonishments
concerning their cultivation practices on the inadequate
land they now hold. The land problem in this degree of
severity in the Federation is unique to Southern Rhodesia.

Africans to the north of the Zambezi, understandably, did not desire that a government which maintained such practices as these should extend its powers over their territories. Actually, even in the Federation they need not have feared this as much as some other consequences. Federation itself implied some autonomy for each entity, certainly on matters as domestic as land tenure. On the other hand, it is clear that Africans elsewhere would sympathize with the Southern Rhodesian Africans's aspirations to possess a larger share of the land. On this basis, objections to the Federation were sound.

The agricultural areas serve some growing cities in both Southern and Northern Rhodesia, and much of the agricultural development in certain regions is directly related to these urban places. Once again, Southern Rhodesia leads the field. Significant urban development may be said to be virtually absent in Nyasaland, and in Northern Rhodesia only Lusaka, the capital, and places on the Copperbelt have attained any size. In Southern Rhodesia, both the capital, Salisbury, and the second city, Bulawayo, and their immediate environs now have a population exceeding 200,-000. Only in Southern Rhodesia has there been a large amount of urbanization of the African population, and as in South Africa the African element of the urban population is larger than the white. The influx has been partly the result of land policies, many Africans being forced to seek opportunity away from their reserve areas. The industries of Southern Rhodesia developed in or near the major towns and cities, drawing Africans to them, and the large white population requiring domestic help has further stimulated the African migration to the growing urban areas.

The industrial development of the Federation has accelerated considerably in recent years. This is virtually to say that Southern Rhodesia's industrial development has undergone acceleration, because the two northern territories contribute little in this field, with the exception of those industries related to the copper mines. Even in Southern Rhodesia, however, industry lagged for a long time. For decades, the riches of the mineral mines and croplands were quite sufficient to provide the necessary cash for imports. Before 1940 there was no real impetus for industrialization, and growth was slow. Then the war brought to Central Africa—as it did to South Africa—a sudden lack of consumer goods. Even imports from South Africa were curtailed, since the Union needed her own production in increasing amounts. This was the first stimulus to industrialization in Southern Rhodesia, and developments since then have been variously described as "sensational," "phenomenal," and unparalleled in Africa and perhaps in the world. This last assertion is in all probability a gross exaggeration, but it is true that the economic situation of Southern Rhodesia changed considerably after 1940. A temporary increase in white population through the arrival of thousands of British and Allied troops served to emphasize the shortage of consumer goods and drove manufacturers to rapid expansion of their production. Secondary industries sprang up, and the industrially dormant territory had come to life.

As the industries developed, the number of Africans and whites employed in them grew, and so did the towns. Among the largest industries is the Rhodesian Iron and Steel Company, which produces a variety of materials used by the railroads and building trades, and in households. Rhodesia's second industry is based on the large tobacco crop and involves the preparation of the locally grown product for home and overseas consumption. Cigarettes are also manufactured in Nyasaland, though the tobacco industry there is on a much smaller scale than it is in Southern Rhodesia. Particularly spectacular has been the progress of the clothing industry, which is now the largest single secondary industry in Southern Rhodesia. This industry satisfies the home market and exports to South Africa and other neighboring territories as well. Together with the textile industry, which produces cotton goods of a different nature such as khaki, calico, canvas, towels, and such, the clothing industry employs over 10,000 people, African and white. There are several large cotton mills at Gatooma, and nearly 150 clothing factories in the various cities and larger towns. Another very important industry is the cement industry, which was in the spotlight during the construction of the huge dam wall at Kariba. With the increase in demand for construction materials, particularly in Southern Rhodesia after World War II, the country became an importer of cement while the local industry grew. Today this growth has caught up with local demand, and production since 1946 has expanded over a dozen times. There are three factories in Southern Rhodesia, and one in Northern Rhodesia and Nyasaland each, and the industry as a whole benefited greatly by the increased demand when Kariba was under construction. It is still able to sell its entire production in the Federation.

There are a large number of additional industrial establishments, among which are sugar refineries, paper-production plants, footwear industries, food-processing factories, furniture-construction firms, and plants turning out soap, beverages, tires, radios, and fertilizer. Trade figures show the decreasing dependence of the population upon imports other than major machinery and vehicle items, although specialized foods have recently shown an increase in the import totals. This is due to the demands of the white settler population of Southern Rhodesia, which is rising rapidly through immigration. Another import which has risen rapidly in connection with economic growth of the area is that of oils, gasoline, and associated products. In advertising the economic progress of the new Federation and encouraging investment, these are facts pointed at by the government, which attempts to create an ever more attractive climate for such activity. Indeed, were it not for the latent political problems of Central Africa, the situation would appear ideal. There is a tremendous, if largely unskilled, labor force in the region, and local markets include not only the wealthy white sector of the Federation but also that of the Republic of South Africa, which has always been closely tied to Southern Rhodesia in terms of trade. In Southern Rhodesia, Africans employed in industry already outnumber whites by ten to one. The number of industrial establishments is rising rapidly, well over a hundred being added every year. About three out of every four such establishments are in Southern Rhodesia, emphasizing further the domination of this territory in the field of industry and economic development. As in South Africa, there is—if less officially—a policy of job restriction, but Africans are intent upon advancement and are willing to work for it. Lately, they have increasingly organized themselves in support of their desires, and in time, they will no doubt come to play more important roles in the country's industries. Already some individuals have managed to attain positions above the unskilled labor level, particularly in the building industry.

| Cities and Towns

Salisbury, Bulawayo, Lusaka, and the other towns of the Federation are very attractive places. After Bulawayo had for some time been the focal point for affairs in Southern Rhodesia, Salisbury's administrative importance and growth soon began to assert themselves, and it became the permanent capital, first of the Colony of Southern Rhodesia, then of the Federation. In 1935 Salisbury was proclaimed a city, but its period of rapid growth did not begin until immediately after World War II, when manufacturing and industry grew apace. In 1962, the estimated population of the capital and its immediate environs is 180,000 Africans, 85,000 whites, and about 4,500 persons of other racial groups.

Salisbury is a city of tall apartment buildings built in a contemporary style which gives the city a real skyline and the appearance of being clean and modern. The city lies amid gently rolling hills in the northern part of the territory and has many wide avenues lined with beautiful jacaranda, bauhinia, and other trees. Fences are lined with bougainvillea, and in Salisbury Park a variety of other local flora can be found. Salisbury is a colorful city with many picturesque spots and pretty suburbs, but it is no dormant provincial town. This city means business and does a lot of it. It has become the Federation's biggest industrial center, and the efficiency of the huge tobacco sales floors is famed. It is a railway center, being connected with Beira on the Moçambique coast, Gwelo and points south, including Bulawayo, and Sinioa and other mining places in the north. There is a project under consideration which would connect Salisbury directly with Lusaka, thereby eliminating the need for the lengthy and cumbersome loop via Livingstone. Should this materialize, then the commanding position of Salisbury will be even stronger.

One of the hubs of the city is Cecil Square, named after Rhodes, where a bronze flagstaff marks, approximately, the site where the Union Jack was first raised when the famous Pioneer Column reached this point in 1890. The Square is the site for flower-sellers, weary shoppers, and a parking problem, but the streets that lead past and into it —Baker, Gordon, and Stanley Avenues—are not unlike those of a medium-sized South African city. In fact, there is much in the bustle of Salisbury, the tall multistory buildings, the rushing shoppers, the jacaranda trees, the large number of cars and extensive suburbs that is reminiscent of South Africa. High-class suburbs form a sharp contrast to the African residential areas—in effect, there is residential racial segregation. Although it is not everywhere and constantly advertised by as many signposts as there are in South Africa, there is no equality here and little interracial living. The difference remains that the door has not been finally closed, and although this does not bring a great deal of satisfaction to the African, who wants equality immediately, there is something of a safety valve in the absence of rigid segregation at every possible point. But the urban areas remain the places where contrasts between white wealth and African poverty, whatever their causes, are strongest and where the tendency to segregate is most obvious. Africans in Salisbury have rioted in an expression of their grievances, and events in Salisbury in 1959 and again in 1960 did much to create the State of Emergency proclaimed here. When in July, 1960, Sir Edgar Whitehead, the Prime Minister, refused to meet with African leaders to discuss African grievances, tension mounted, and repeated rioting took place in which whites and Africans were involved. Measures introduced to prevent riotous assembly strongly resembled those passed for similar reasons in South Africa. Salisbury's trouble spots are Highfields and Harare, suburbs where a large number of the African residents of the city live. Serious violence arose for such trivial reasons as a minor road accident involving a white driver and an African pedestrian, a bar refusing to serve an African (though other Africans were being served) because of his incomplete attire. What occurred was a clear indication that Salisbury was and is a place of tension. The tension has fundamental causes, among which unemployment is not the least. Although industries are growing, there are always more Africans coming to the cities than can quickly find work. South Africa has sought to combat the problem of unemployment with pass laws. In Southern Rhodesia, there are those who favor a similar course of action, and in fact there is already a trend in this direction.

Bulawayo, in the southwestern part of Southern Rhodesia, is the Federation's second city. Not only does it vie with Salisbury for domination in the country, but it is more than twice as large as its next rival. Of the nearly 200,000 people in the city, about 140,000 are Africans and over 50,000 are white. "Bulawayo" means "killed" or "place of killing" in an African language, and here it was that Lobengula, the last of the Matabele chiefs, had his capital. Bulawayo, like Salisbury, has enjoyed a period of very rapid growth. It, too, is today a city with a modern appearance, tall buildings in the central business district, wide avenues, and magnificent flamboyant trees. Main Street, a busy, wide thoroughfare with diagonal parking on both sides and in the center, is graced by an imposing statue of Rhodes at the center of the busy intersection adjacent to the post office. Beyond the central city are extensive suburbs as spacious as any, and the African housing areas are said to rank among the best in all Southern Africa. The high total of African population results from the considerable industrial development of Bulawayo city and its surrounding region. Here is the Federation's greatest concentration of engineering and metal industries, and the clothing and textile industries have grown here more rapidly than any other place in the country.

Bulawayo serves as the threshold to the Federation for all communications and transport from the south. Through this center pass all products arriving from the Republic of South Africa. It is directly linked with the port of Lourenço Marques, and so long as the projected Salisbury—Lusaka railroad is uncompleted, Bulawayo remains nearest to the Copperbelt by rail. The city also lies close to the great coal fields of Wankie and the iron deposits at Que Que, and it is the center for some of the Federation's best cattle country. Not surprisingly, Bulawayo gets a large share of the electricity supply from Kariba. In the immediate vicinity of the city lie mines producing sixteen different minerals which together make up, in value, fully half the annual Southern Rhodesian production. Gold and tungsten figure prominently, as do asbestos, coal, and tin, all mined within Bulawayo District. It is not surprising that s significant town developed here, and that this town soon reached the proportions of a city. Bulawayo has had the honor of being a city since 1943.

A number of smaller towns have been established in various places in Southern Rhodesia. They all serve their surrounding region, and several are thriving. Umtali is on the railroad and road from Beira and lies near the Moçambique border. This beautifully situated town has a certain amount of industrial development and serves as a distribution center. Not surprisingly, a number of the other towns lie between Salisbury and Bulawayo, not far from the Great Dyke, and on the "ridge" of the plateau. These include Gwelo, Que Que, and Gatooma, the latter an important center of the cotton industry. All these towns, however, have a population of less than 30,000.

Northern Rhodesia cannot compete with the south in terms of urban growth. Situated on the railroad which connects the Copperbelt via picturesque Livingstone (the former capital of the north) to Southern Rhodesia, is Lusaka. This town owes its origin to the discovery here of a large limestone outcrop by an individual involved in the construction of the railroad. Its initial growth was slow, but its favorable location resulted in its being selected as the new capital of the territory after the assumption of administration by the British government. In 1935, preparations having been in progress for some years, the transfer took place, and Livingstone ceased to be the seat of administration. Since then, Lusaka has not had a period of spectacular growth, but particularly during recent years, the rate of progress has increased. Since 1951, the white population has trebled from just over 4,000 to well over 13,000, while the African population increased from 25,-000 to nearly 80,000. No town in Northern Rhodesia grows faster, but as the totals reveal, there is no competition here for Salisbury and Bulawayo. There are some fine new buildings on Cairo Road, and the town shows signs of expansion on a bigger scale than ever before. But although located on a major railroad and at a junction of several main roads, the comparative remoteness of Lusaka is always likely to inhibit growth. Of course, changes can be brought about, such as by a new railroad linking the city to Salisbury, the provision of abundant cheap power from the Kafue River, or the success of a large-scale agricultural scheme in the Kafue Flats. Meanwhile, Lusaka's main economic asset is its location in the center of a large farming community. There is a modern and well-equipped tobacco grading warehouse, and an annual agricultural show is held. Lusaka may have a bright future, if any one of several possibilities materialize.

Northern Rhodesia's other towns lie on the Copperbelt, and they are primarily mining towns, rather large ones, considering their single purpose. Kitwe rivals Lusaka in size, and Ndola is likewise a most important center. Ndola, at the very end of the railroad in Northern Rhodesia, is not only a mining center but also has industries and performs a distributing function for much of the Copperbelt. The characteristic of many of the Copperbelt towns is the limited white population, usually less than 10,000, and the large African sector, generally over 50,000. In the Copperbelt, much has been achieved by active trade unions working for the advancement of the African worker. Although there has been resistance on the part of many whites, there is a great deal of integration in the Copperbelt towns, and Africans have advanced into better positions than they were formerly able to hold. Thanks to government action and the cooperation of some admirable individuals among the white population of the Copperbelt, there are now unsegregated hotels and theaters—not universally, but they are there, and they function well. This is a considerable achievement, and although it would be incorrect to assume that non-segregation and non-discrimination are the rule, the progress made in the Copperbelt in terms of race relations must not be ignored.

Nyasaland's urban development is negligible, an indication of the poverty of the region. The capital of the territory is Zomba, which lies at the foot of Zomba Mountain in a most picturesque situation. Zomba's total population, however, can be counted quickly. There are less than 1,000 whites in this town, mainly involved in the administration of the country, and perhaps 8,000 Africans in employment. The area of Blantyre-Limbe, located on the railroad, is perhaps the closest approach to an urban center with some economic justification. There are about 70,000 Africans and nearly 4,000 whites, and the town boasts two tobacco auction floors. It deserves meditation that the only other urban agglomeration is in the Central Province at Lilongwe, which has a population of about 15,000 Africans and a few hundred whites. Here is a country with some 3 million people and not a single town of 100,000 to boast of.

Again, the differences to the north and south of the Zambezi are very obvious. Southern Rhodesia has the benefits as well as the evils of urbanization on at least a moderate scale. Neither Northern Rhodesia nor Nyasaland possess this manifestation of progress to a comparable degree.

Educational and cultural institutions and organizations are largely confined to the cities and thus mostly to Southern Rhodesia. Although the government is building several African schools, the great majority of African children still go to mission schools, which are government-subsidized. For whites, there are a number of good schools in each of the territories. In Bulawayo, there is a teacher's college, and the University College of Rhodesia and Nyasaland opened its—unsegregated—doors in 1957. The University College is situated in Salisbury and serves a wide area. There are no professional theatrical companies even in the larger cities, but Southern Rhodesia is usually included in the itinerary of artists visiting Southern Africa. There are semi-amateur orchestras and amateur theater groups, and Southern Rhodesia perhaps has the excuse which South Africa lacks: it does not have an adequate population to sustain professionals. In Salisbury are the headquarters of the Federal Broadcasting Corporation, which puts out a daily program based on the B.B.C. interspersed with commercially maintained broadcasts. The commercial and non-commercial programs will be separated and parallel in the future. There are regional program studios in Bulawayo and Kitwe, as well as Lusaka and Blantyre, the last two being African studios, preparing programs in six African languages as well as English. At the present time, the Federation is developing television as an additional medium of communication, which is rather strong evidence against the South African argument that in the Transvaal this is not economically feasible.

Two of the major newspapers published in the Federation are the *Rhodesia Herald*, a daily printed in Salisbury and read mainly by whites, and the *African Daily News*, also produced in the capital but circulated mainly among

Africans. Bulawayo reads the *Chronicle,* and Ndola the *Northern News.* In addition, there are several dozen week-lies, semi-dailies and other periodicals, of which a number are printed in African languages as well as English. The *Herald* and Salisbury's other daily, the *Evening Standard,* are comparable to the better English-language newspapers of South Africa, and the quality of the press may be said to be good. Circulation is not very great. Competition is not as keen as it is in South Africa's major cities, and generally the newspapers play a straight reporting role rather than that of political party organs. Sales are almost confined to the cities or towns where publication takes place and those along the railroad. Only some of the African weeklies seem to be spread—by being passed from hand to hand—throughout the country. With the present communication problems, the Federation cannot be expected to have a na-tional daily paper. The *Herald* (like South Africa's *Star*) comes nearest to serving such a function. As in South Af-rica, however, the cities dominate their surrounding areas to such an extent and are still so isolated in their dominat-ing position that inter-city competition in the field of news communication is practically absent. Much water will flow over Kariba before this situation is changed.

| Politics and Government

The accelerated economic development of the Rhodesian region has brought with it an increase in the speed of political evolution. Northern Rhodesia, after the Char-tered Company had handed over its administration to Lon-don, became the scene of the great mining boom which led to the development of the Copperbelt complex. Until 1924, the territory had been backward and poor, but the Com-pany, now freed of its governing duties, embarked upon a policy of vigorous economic development. As a result, whites were entering the territory faster by 1930 than ever before, and although the immigration rate never reached the proportions it did in Southern Rhodesia, the nucleus of a white-settler population began to form. This new body of immigrants had economic motives for their entry, and they became a new factor in the political development of the territory. They were unable to alienate much land, and they settled in clusters along the railroad and in the mining towns. But unlike their predecessors, who had mainly been administrators, missionaries, traders, and hunters, these people were there to exploit the resources of the country. However, political organization among the new Northern Rhodesians was slow in coming, and African chiefs and headmen continued to gain administrative powers. In Nyasaland, a similar situation existed. In 1929, Africans were given the right to preside over African courts, a system which was entrenched in 1933 through legislation. Africans also obtained some administrative jurisdiction in Nyasaland, where a system of treasuries run by Africans was established. Most important of all, a mechanism for local government had, through this and other moves, been set up.

South of the Zambezi, there was no such legislative action and no such mechanism. Native Commissioners retained the judicial and other powers, and the Land Apportionment Act contrasted sharply to the protection of land rights of Africans in the north. The depression of the early 1930's came at a time when all the adjectives now used to describe the Southern Rhodesian economy did not yet apply, and it had disastrous effects. It also stimulated political division, particularly in Southern Rhodesia. This period was marked by the rise of the Reform Party, led by Godfrey Huggins, who was later to become Lord Malvern, the first Prime Minister (1953–56) of the new Federation. The Reform Party was a white man's party, which wanted protection for white workers against African competition and which aimed at the removal of Africans from the common voters' rolls, among other demands. In 1933, the Reform Party won the election, and Huggins became— and was to be for the next twenty years—Prime Minister of Southern Rhodesia. The Reform Party's rival at this time was the Labor party, which initially was strong, but a new political crisis occurred within a year. Huggins formed the United Party and in the general election that followed, this new group won a notable victory. The total Opposition amounted to 6 seats out of 30 in the single-chamber Parliament. These events were followed by economic recovery, and Huggins and his governing party went unchallenged. Immigration began to expand, and the country developed along South African lines.

In Northern Rhodesia, meanwhile, the depression, coming after a considerable amount of white immigration, was also having political consequences, even though in 1930 the British government through its Secretary of State reiterated that African interests should remain paramount. Local whites began to feel concern about the increase in the Africans' powers, even if this was largely confined to African areas. They realized that in time Africans might come to rule the entire country, and they began to look toward the south, where it was the whites who were gaining in strength. The transfer of the capital from Livingstone to Lusaka was resented particularly in Livingstone and Ndola. All these and other considerations brought about a conference on the question of amalgamation between the Rhodesias north and south of the Zambezi, and this was held in 1936 at Victoria Falls. Southern Rhodesians had come to look upon Northern Rhodesia in a manner very different from that of 20 years earlier. Now a thriving territory with a growing mineral output and increasing white settlement, the north was an asset rather than a liability in the eyes of the settlers south of the Zambezi. The 1936 conference of local representatives, not surprisingly, reported in favor of joining the two Rhodesias, and in 1938 the British government sent a Royal Commission under the direction of Bledisloe to investigate the matter. Although Rhodesians like to pretend that the Bledisloe Commission reported in favor of amalgamation, it actually concluded that the divergent racial policies to the north and south of the Zambezi made such a procedure impracticable. The Commission indicated that although closer ties between the territories were desirable, they had best be confined to closer scientific and judicial cooperation.

Thus while the whites of South Africa were appealing to London for the transfer of the High Commission Territories, the settlers in the Rhodesias wished to increase their power by spreading white domination as practiced in Southern Rhodesia to the north of the Zambezi River. It is another in a series of remarkable parallels, with demands for protection of white workers, removal of Africans from the joint voters' rolls, land alienation, and laws to separate white and African land all coming during this period in Southern Rhodesia as well as in the Union of South Africa. And as in the case of the negotiations concerning the High Commission Territories, several matters—in the Rhodesias particularly the question of amalgamation—ceased to be discussed when World War II broke out in 1939.

The war had its economic and political effects in the Rhodesias as elsewhere. Northern Rhodesia's capacity to produce copper, a vital mineral, was greatly increased, and there was no collapse after the war effort was over, the country permanently ranking among the world's top producers. The war also stimulated rapid industrialization in Southern Rhodesia, where secondary industries sprouted everywhere. Here, the United and Labor Parties formed a coalition, with Huggins continuing on as Prime Minister. The economic policies—such as the taking over of the growing iron and steel industry by the government—led to a split and the formation of the Liberal Party, which became the Opposition. In Northern Rhodesia, Roy Welenssky, a former professional boxer and train engineer, gained prominence at the head of the successful Northern Rhodesian Labor Party, a white organization. The rapidity of expansion caused labor trouble and increasing friction over the question of which skilled jobs should be allotted to white people and which to Africans. This, the closer contact among Africans and whites, was one important effect of the economic progress in the Rhodesias in general. In Southern Rhodesia the flood of Africans to the cities had

really begun, and it soon became obvious even to segrega-
tion-minded Southern Rhodesians like Huggins that some
amelioration of the policy of separate development was
necessary. In Northern Rhodesia and even in remote Nyasaland, least affected by the war's consequences, important
constitutional changes took place. African Regional Councils were formed to advise the government on matters of
local as well as territorial importance, and in 1948, Africans in Northern Rhodesia gained admission to the Legislative Council. Soon, beside the 10 white members of this
Council representing white interests and two whites representing African interests (and who were nominated), there
were 9 African official members and 2 African "unofficial"
members. In Nyasaland, also, Africans were making progress in the political sphere. They were placed on the Legislative Council in 1949, although in very small numbers (2
out of a total of 19 members), when the Asiatics in the territory also were given representation through one nominated member. These steps fostered the development of
the African nationalism which has in recent years become
such a vital factor in the Federation. Africans began to
organize themselves politically. The Nyasaland African
Congress formed in 1944. In Northern Rhodesia a number
of societies eventually jelled into the Northern Rhodesia
Congress in 1948. In 1949, the African Mineworkers'
Trade Union was created. As the Africans showed signs of
political consciousness and began to demand ever greater
powers in their country, whites to the north of the Zambezi
began once again to contemplate union with Southern Rhodesia. Huggins had long been known to favor this amalgamation, and the economic factor in the desire for federation grew stronger than ever. The Copperbelt needed
electric power and more capital, and an economy tied to
Southern Rhodesia's promised to bring these and other
benefits. Hence agitation for amalgamation was renewed.

In Southern Rhodesia, however, the political situation
had changed, for in 1946 Huggins had barely squeezed
past his new Liberal Party opponents, gaining just 13 seats

against 12 for the Opposition. The small Labor Party held the balance. Southern Rhodesians turned all their attention to internal politics. The economic boom after the war, however, boosted the government considerably, and the 1948 Nationalist victory in South Africa had much impact here. Late in 1948 Huggins and his United Party, probably as a result of these two factors, gained an overwhelming majority at the polls, defeating the Liberal party by 24 seats to 5. The Labor Party lost all but one seat. It was only after all this had come to pass that the question of federation between the Rhodesias and Nyasaland could finally again be studied.

Thus 13 years after the 1936 meeting on amalgamation held at Victoria Falls, another such conference was held at the same site. There were representatives from the Rhodesias as well as Nyasaland, and when the possibility of setting up a unitary state seemed remote, federation plans were laid. The two major figures at this conference were Huggins from the south and Welensky, still the leader of whites in the north. Pressure was put upon the British (Labor) government to appoint another commission of investigation, and this time, unlike the Bledisloe Commission, it reported in favor of federation. To many, this was a great surprise, and in Northern Rhodesia and Nyasaland, where Africans were politically organized and could express their opinions, there was sustained and undivided opposition to the scheme. To them, rule from Britain was cherished as a protection against local white domination as it existed in South Africa (where a similar decision for union in 1910 had deprived Cape Africans of their rights) and in Southern Rhodesia. Neither were all whites in the Rhodesias in favor of federation. Extremists on one side saw African rights increased in the south and abhorred the idea that Africans should sit next to whites in a Federation Parliament. On the other side were those who sympathized with African fears that federation would spread white domination across the north. They were few, but the philosophy did exist. Finally, a number of citizens felt that if Southern

Rhodesia was to federate with any country, it should be South Africa.

As so often in the history of Africa, it was an occurrence in Europe that shaped the course of events. The Labor government in Britain, which did not sympathize with those who favored federation, after investigation decided to hold the matter in abeyance. In 1951, this government was defeated by the Conservatives, who immediately announced that they favored the scheme and intended to carry it forward. Bitter debates occurred as the Labor Party Opposition now made it official policy to oppose federation. Nevertheless, a referendum was organized in Southern Rhodesia. Here, qualifications for being placed upon the voters' rolls were financial rather than strictly racial; to become a voter, an adult had to have property valued at $1,400 or an annual income exceeding $672 and must pass a simple test in the English language. These measures virtually eliminated the African from the voters' rolls in any numbers; an estimate of the electorate at the time of the referendum suggests that of 50,000 voters in Southern Rhodesia, perhaps 400 were Africans. In 1951, a last effort to entirely eliminate the Africans from the voters' rolls had failed, largely because of strong disapproval in London. It was then that the financial requirements for voter status were raised drastically, with the supporting speech in Parliament stating bluntly that this was a scheme to reduce African voter representation. It had the desired effect. The referendum, held in the first half of 1953, approved federation by a large majority. In the legislatures of all three territories, the plan was likewise endorsed. In Nyasaland and Northern Rhodesia, there were angry scenes of bitter reproach, as Africans saw their progress toward self-government come to a halt. Never before were relations between white and African as strained. In Southern Rhodesia, there were few channels through which Africans could express their opinions in the matter, since they were politically less organized.

Thus, on September 3, 1953, the new Federal State of

Rhodesia and Nyasaland was inaugurated. A Federal Assembly was established, where 26 of the 35 members were whites representing the interests of about 200,000 settlers, and the remaining 9 (6 Africans and 3 whites) represented those of 7 million Africans. The individual territorial governments retained responsibility for local administration. At the end of 1953, the first federal election was held. The supporters of Huggins (in Southern Rhodesia) and Welensky (in Northern Rhodesia) had formed the Federal Party. Huggins had changed his views considerably since his Reform Party days, when he favored South African–type segregation. Now, he was an advocate of greater interracial cooperation. His Federal Party completely overwhelmed the opposition, the Confederate Party. This party consisted mainly of people opposed to federation, and although it got nearly a third of all votes cast, it gained only one seat and was rendered entirely ineffective. The Confederate Party, since this first election, has changed its name and leadership several times. Huggins became the first Federal Prime Minister and was subsequently given the title of Lord Malvern, while his adjutant, Welensky, became Vice-Premier. Garfield Todd became the new Prime Minister of Southern Rhodesia.

Salisbury was chosen for the capital of the new Federation, with strong competition from Livingstone and Lusaka giving rise to a prolonged and intense debate in the very beginning of the first parliamentary session. Soon, however, more fundamentally important issues were to be discussed by the new Assembly. One of the provisions of federation had been the establishment of an African Affairs Board, which could review any action taken by the Assembly with reference to its consequences for Africans. If such action were held to be discriminatory, then the British government would review and possibly veto the legislation in question. Among the political moves which occurred within the first five years of federation were the Constitution Amendment Act and the Electoral Bill. By then, Sir Roy Welensky had taken over from the retired Lord Malvern, and the Federation appeared to be drifting in the direction of racial trouble. The Constitution Amendment Act had as its main objective the enlargement of the Federal Assembly from 35 members (9 African representatives) to 59 (15 African representatives). The Electoral Bill involved sweeping changes in the requirements for the franchise. An "ordinary" and a "special" voters' roll were created, with the voters on the "ordinary" roll electing white members of the Assembly and then combining with the voters on the "special" roll to elect African members and the one white member who represents African interests. In effect, it was a racial separation of the voters' rolls or nearly so. The "ordinary" roll requires an annual income of over $2,000 or ownership of property worth at least $4,200 or a series of alternate qualifications which could be met by only a handful of Africans. The "special" roll, on the other hand, requires much less in the way of annual income, and this was the roll on which Africans could register in numbers. In addition, a series of elaborate checks and balances was devised, whereby the weight of the vote was kept in favor of the "ordinary" roll. It was, actually, an extension of the 1951 provisions, and though there are no racial qualifications, as such, for the separate rolls, the effect is not very different from separate registration. The African Affairs Board held both measures—the Constitutional Amendment Act and the Electoral Bill—to be discriminatory, but when the British government considered them, they were approved.

Thus ended the first term of the Federal Party government. In 1958, Welensky campaigned on the basis of these achievements among others, and his opponents formed the Dominion Party. The election was watched with great interest by all students of the African scene, particularly in view of the success which the Dominion Party had just had in the previous Southern Rhodesian local elections. Welensky, however, led his Federal Party to another great victory, gaining 46 seats, while the Dominion Party secured only 8. One seat was won by an independent politician, and the remaining seats were those of the "specially elected" African members and appointed white members for Northern Rhodesia and Nyasaland. One effect of this very complicated machinery is that African members are found on both the Government and Opposition side of the Assembly, so that there is not an absolute division on racial lines— although the African members, particularly those from the north, are implacable opponents of everything the government proposes, largely because they are opponents of the Federation itself. One of the consequences of the sweeping victory of the Welensky forces and the new numerical arrangements in the Assembly was that the African Affairs Board now had a pro-government majority, making it, to say the very least, highly unlikely that it would continue to describe government-sponsored measures as "discriminatory." Welensky must have been as eager to terminate the effectiveness of the Board as Malan was to end the powers of the Supreme Court.

African opposition to the Federation has increased rather than waned, and events in the Federal Assembly have much to do with this attitude. The whole basis for this aversion has been misinterpreted as sheer ignorance, lack of willingness to cooperate, and deliberate sabotage of everything done by white men. African leaders in the north, however, cannot fail to note that the same people who maintain the discriminatory policies in Southern Rhodesia are the people who elect the bulk of the Federal Assembly, and that the Electoral Bill has given whites the domination of the electorate for many years to come. Lord Malvern, reflecting on the first five years of the Federation, wrote in *Optima* in 1959: "Africans were not accustomed to make any decisions other than when and where the next beer drink was to be, how they were going to pay for a wife or what the crops would be like. . . . Unfortunately, when federation was discussed in Nyasaland, it was laid down that Africans must be given an opportunity of making up their own minds. . . ." This is a highly significant comment, lest it be forgotten that irrationality in Africa is displayed by white as well as black. In this same paper Lord Malvern states that the Nyasaland and Northern Rhodesia African National Congress was a body "with a grand name but which, at the coming of federation, represented, at most, about five per cent of the native African people. . . ." Five per cent of the African population of Northern Rhodesia and Nyasaland totals about 250,000 persons, or a good deal more than all whites in Southern Rhodesia and several times the entire white electorate!

Welensky's 1958 victory over the Dominion Party was, in a way, evidence of some moderation on the part of whites in the Federation. The Dominion Party stood for immediate independence from Britain, its leaders sounding rather like Nationalists further south. Welensky's own aim is likewise a revision of the constitution and greater independence, but through negotiations. Briefly on the scene in Southern Rhodesia was the moderately liberal Garfield Todd, whose government lasted from 1954, when his United (Federal) party defeated the Opposition, to 1958, when a crisis broke up his cabinet. Todd resigned, and Sir Edgar Whitehead replaced him as Prime Minister of the territory. In the ensuing election, Whitehead defeated the Dominion Party somewhat narrowly by 17 seats to 13, and Todd and his liberals, whom he had gathered in a separate party, failed to win a single seat. There was, therefore, a straight division now between the Welensky-Whitehead people, who were conservative, and the ultra-conservative Dominion Party, whose stand for immediate independence, segregation, and white domination resembled that of the government in South Africa. It cannot be a surprise that Africans, particularly in the north where they had in their own Councils some means of expression, did not throw their hearty support behind the Federation effort. Nationalism was rife, and calls for secession began to be made. However, as in South Africa, Africans were not united in their demands. Those who sat with the white legislators were called traitors when they refused to give up their seats in a display of protest. These people, who refused to resign, were as opposed to the government as were the Africans who back home demanded their resignation, but they felt that they could do more good in the Assembly than outside of it. A great rift developed, and even the return in 1958 of Dr. Hastings Banda, the Nyasaland leader of African nationalists, failed to end the bitterness. In Northern Rhodesia, Kenneth Kaunda broke away from the main Congress to form his own splinter group, and the weakness of the African organizations were open for all to see. However, it remains a great danger to mistake the relatively small size and fragmentation of the African movements for a lack of common purpose. There would appear to be a great deal more understanding of the basic issues at hand than many whites will admit, and the mood of the Africans in Nyasaland and Northern Rhodesia is unmistakable. They may not have known much about the meaning of the Federation when it was established without their consent. Today, they are mostly determined to break the Federation up and to form other associations with African nations to the north and east—somewhat unrealistically, perhaps, but nationalism has never been blessed with a great deal of rationale.

Meanwhile, Welensky is pressing for constitutional reform and for greater independence—which means, of course, more jurisdiction for the elected government in the Federation, including the north. Lord Malvern in 1956 failed to make progress in this sphere, his plans having been rejected in London. Welensky in 1957 secured the promise that a conference on constitutional reform would be held between 1960 and 1962. This conference took place in December, 1960, and it was attended by, besides representatives from the British government, Welensky, Whitehead, Banda, Kaunda, and Nkomo, the last being the leader of the Southern Rhodesia United National Democratic Party. The three Africans walked out after the first week, since the white representatives did not wish to admit the possibility of an African-elected majority. However, the British government, though still Conservative, has been somewhat less anxious to grant the Federal government its every wish, and after the conference, which ended in utter failure, the Commonwealth Relations State Secretary indicated that the British government is now set upon a three-point policy: there must be white and African "partnership" (a term which means many different things to different people in Africa); there will be no dissolution of the Federation or any other move that might hamper economic development; and Africans will have to play a bigger role in Federal affairs.

Recently, the Federation has been the scene of some violent reaction to political and racial friction which have for long been building tensions. Banda's return was accompanied by rioting in Nyasaland which had clear racial overtones—this where Nyasaland has for so long been a place of interracial cooperation and harmonious progress. The Southern Rhodesia African National Congress was banned, leaders were arrested, and a state of emergency was proclaimed in the territory. A similar course of action was taken in Nyasaland, and in Northern Rhodesia the African National Congress was also banned. Rather than increasing cooperation, the races were drifting farther apart, with little hope for future harmony. The tragedy is that some progress really had been made toward peaceful political evolution. On the other hand, there is still hope in the Federation. There is no absolute territorial segregation, and Africans and whites are still sitting in the same Assembly chamber. The Federation itself may not survive, but in Central Africa it is not the most important issue at stake. At all costs, Africans must not be hurried into rejecting liberal and moderate whites and thus turning to untempered racialism.

Britain's reluctance to provide greater powers to the Federal government has been reassuring to some African leaders, while others now want nothing less than the fragmentation of the Federation and the complete autonomy of their own territory and are dissatisfied so long as this has not come about. There are extremists among the Africans just as there are among the whites, and they are gaining support. There were whites who, upon the rejection in London of demands for increased powers, threatened to effect immediate independence by going to war over the matter, to fight a "war of independence," one or two references being made at the time to the similar battle fought by the American colonies. The powers that were so badly wanted include more direct local administration of affairs in Northern Rhodesia and Nyasaland. While there was not a great change in the social scene in Southern Rhodesia when the Federation materialized, it was hoped in Salisbury that British rule in Northern Rhodesia and Nyasaland would be replaced, through constitutional development, by greater local government, which could, through manipulation, be dominated by the white minority. Welensky and his followers feared that Britain would continue to develop the two protectorates in the direction of majority rule, which would almost certainly mean the end of their participation in the Federation. Although the Africans had not been heard when Federation had been discussed, they were now more vocal, threatening Mau Mau–type violence, and more militantly nationalistic. The British government, and particularly the Secretary of State for the Colonies, Macleod, faced a most difficult situation. Having helped create the Federation itself, the London government faced a gathering storm over the behavior of its rulers, who were more and more frequently compared to South Africa's Nationalists. Meanwhile, Ghana, Nigeria, and other African states had begun to exert political influence, taking a stand of hostility to the Federation, of course. While wishing to keep its actions with respect to Ghana and other former colonies compatible with those in the Federation, the British government was here dealing with a white people who wanted autonomy. Southern Rhodesians argued that Ghana, no more a democracy than the Federation, had become an independent state because its rulers were African, while in the case of the Federation a similar request was denied because those who demanded independence were white. The African nationalists, on the other hand, reminded the British government of their opposition to the extension of powers for the Federal government, and whatever move was made in London, it was sure to cause violent reaction in Central Africa. The divergent nature of the Federation north and south of the Zambezi had come to the fore, and the dangers against which the Bledisloe Commission had warned in 1938 were now having effect. The Federation's economy, which was supposed to benefit from political unity, now began to level off. Measures had to be taken to prevent the outflow of capital.

In London, constitutional talks were held for Northern Rhodesia and Nyasaland immediately after the collapse of the Federal constitutional talks. One of the results of these talks was the renewed assertion on the part of the British that protection would remain so long as the African people desired it. Another affected the franchise so as to increase African participation in elections by perhaps two thousand. A general increase in African participation in local government was envisaged. Even these mild recommendations infuriated many white Rhodesians. Angry anti-British protest meetings occurred early in 1961, and troops were called out to maintain order. Renewed calls for a fight for independence were made, and relations between the Federalists and the British stretched to the breaking point. One of Welensky's main arguments was that Southern-Rhodesian whites had shown good faith by accepting "partnership" and agreeing to sit with Africans in the Assembly. Whatever their race, people in the government must be competent, said Welensky, and they must be people who prevail in the economic, technical, business, skilled-labor, and administrative spheres. By insisting upon the "best qualified" to be the governing people in the Federation, Welensky automatically thereby demanded that whites should everywhere retain the majority. The Northern Rhodesia constitutional decisions were referred to as "feeding Northern Rhodesia to the crocodile of extremism." African advancement in partnership was not seen as impossible, but should proceed more slowly. As long as

African nationalists uttered threats of violence and perpetrated acts of intimidation, Welensky concluded, they
were setting back the clock of their own progress. In reality, the whites were attempting to hang on to such privileges and advantages as they have enjoyed for many decades, and for the first time in many years they had to do
battle to retain what they had. Individual statements threw
much light upon the mentality of many. Lord Malvern,
taking an active interest in the Federal state which he had
been the first to lead, announced that "the British form of
democracy is not for export to Africa." This, while many
Africans were enjoying its benefits and learning its lessons
while studying in Britain! Mr. Macleod was described in
Lusaka as a "lackey" of the "stupid and incompetent" British government, and Sir John Moffat as "a white kaffir."
These outcries, the racial rioting in the cities, the arrest of
African leaders, the banning of African political organizations, and the declaration of the state of emergency all
evidenced just how severe the rift between black and white,
north and south was becoming.

In Africa, the long-obeyed maxim to "make haste slowly"
no longer applies, and its consequences are now a harvest
of reproach and bitterness. Whites in the Federation would
like to retain the status quo as long as possible, and so
would whites in Overseas Portugal and South Africa. But
the Wind of Change has already made its inroads into the
south, and it is disrupting the Federation, the core of the
buffer zone. For many years, haste has been made slowly in
the Rhodesias. Now that some Africans have acquired education, skills, and means, they observe their own people
lacking and backward, and they see an opportunity to wield
power through their own competence and their ability to
sway the support of the masses behind them. They cannot
be expected to delay their own progress, and in fact they
have themselves had trouble keeping their followers in
check. Neither is there now time to educate the masses
which are so willing to fall in behind a leader who may not
really have their own interests at heart. Whites in today's

Africa may well, while they build monuments in praise of their ancestors, occasionally remind themselves that their forefathers made great errors of omission, which are now having dire consequences. Cooperation and partnership could have been possible, had those who made so much profit in the exploitation of Africa stopped to plough some of this wealth back in the form of education. Today, the division within the Federation is almost hopeless, and the future of the Federal State is precarious. The African has ceased to trust his white countryman and suspects his every political move. Making haste slowly in Africa was a profitable philosophy over many decades. These years are over now. Many of the whites in Africa so often say that they, too, are Africans and want to be recognized as such. The "white Africans" who really matter in the Africa of today are those who are willing to accept a temporary setback in status, income, and privilege, while staying on the continent in an effort to aid the African in his new venture of dominant participation in government. Africans are no more trustworthy, able, persevering, and responsible than are whites. Like the whites, they have their own interests at heart, and it would be unrealistic to expect them now to take over as models of democratic, fair, and competent administrators. They have not had much chance to learn, and they have not always had good examples to imitate.

As the Federation faces the Wind of Change, it constitutes the last barrier between South Africa's complete white supremacy and the independent African states. The Zambezi marks the line of division within the Federation itself between advancement for Africans, as in Northern Rhodesia and Nyasaland, and their very limited participation in government and virtual social segregation, as in Southern Rhodesia. The Federation, as might be expected in a buffer zone, has the characteristics of both "black" and "white" Africa. Today, it is the focal point of change in fast-changing Africa. Should the Federation fracture and Nyasaland and Northern Rhodesia gain independence or associate themselves with another Federation such as that of East Africa, then the Zambezi will once again play a role in dividing Southern Africa into two opposites. Once, it was here that white conquest ended and peaceful penetration began. Then, the Wind of Change was blowing northward from the Cape. Now, it is toward the south that the wind blows, and African independence may reach the great river, flowing slowly today through a region of ferment.

FURTHER READING

FRANCK, T. M. *Race and Nationalism: the Struggle for Power in Rhodesia—Nyasaland.* Fordham University Press: New York, 1960.

HANNA, A. J. *The Story of the Rhodesias and Nyasaland.* Faber & Faber: London, 1960.

JONES, A. C. *African Challenge: the Fallacy of Federation.* Africa Bureau: London, 1952.

LORD MALVERN. "Must Nyasaland Be Crucified by World Ignorance?" *Optima,* Vol. 9, No. 4, 1959.

MASON, P. *The Birth of a Dilemma: The Conquest and Settlement of Rhodesia.* Oxford University Press: London, 1959.

STONEHOUSE, J. *Prohibited Immigrant.* Bodley Head: London, 1960.

VIII

From Colony to
Province:
Portugal in Africa

O f all the colonial powers, Portugal alone in 1961 still faced the Wind of Change without having suffered any major loss of territory. Long ago, the Dutch were ousted from Indonesia, but the Portuguese retained their half of the Indonesian island of Timor. Britain withdrew from India, but Portugal remained in Goa longer, despite pressure. France and Britain freed virtually all of West Africa, but the Portuguese stayed in their Guinea. Belgium was driven from the Congo, but Cabinda, the tiny Portuguese enclave on the Congo River mouth, was not affected. For years it seemed that the Portuguese possessed the magic formula, that they were the ideal colonizers, whose non-racial attitudes prevented any friction from developing and who would integrate their overseas territories successfully into a Greater Portugal, this coming as a voluntary rather than a compulsory arrangement. While news of savage racial violence emanated from Kenya and the Congo was rocked by unprecedented political unheaval, Angola and Moçambique, the two large Portuguese territories in Southern Africa, were quiet and, from all appearances, set upon a peaceful course toward political maturity.

Angola and Moçambique are extremely old dependencies, their coastlines having been settled by the Portuguese nearly five centuries ago. Although the Portuguese were briefly ousted by other powers—from Angola by the Dutch and from part of Moçambique by the Arabs—they have held virtually uninterrupted sway over the regions they settled so long ago. Actually, the hinterlands in the interior over which the territories now extend were claimed only during the second half of the 19th century, and effective rule here is less than a century old. But through trade and conquest, the Portuguese have been the masters over the regions they now hold for longer than any other European power in any other part of subsaharan Africa. The Portuguese were the first to travel up the Zambezi, and they probably saw Lake Nyasa long before Livingstone was born. It is likely that they also discovered Zimbabwe, and in both Moçambique and Angola, they long carried on the slave trade, displaying a reluctance to terminate it. However, their effort to obtain the intervening area of the plateau, presently occupied by the Federation, was foiled by the British, who established themselves here and fought Portuguese and Arab slave-traders in what is today Nyasaland. When the boundaries of subsaharan Africa had been finally delimited, Portugal found itself with over 481,000 square miles of Angola and nearly 300,000 square miles of Moçambique, the two territories together possessing a coastline of over 2,200 miles. They literally form the two flanks of the plateau which the Federation of the Rhodesias and Nyasaland covers. Along the coastline lie the natural harbors for the plateau interior, and virtually all exports from the Federation today pass through Portuguese African ports. Moçambique and Angola were occupied by Africans who for long resisted Portuguese domination but were eventually subdued. The African population of smaller Moçambique is somewhat larger than that of Angola. There are over 6 million people in this prorupted eastern territory, and a very heavy concentration in the south, along the coastal regions to the north of the capital, Lourenço Marques. An-

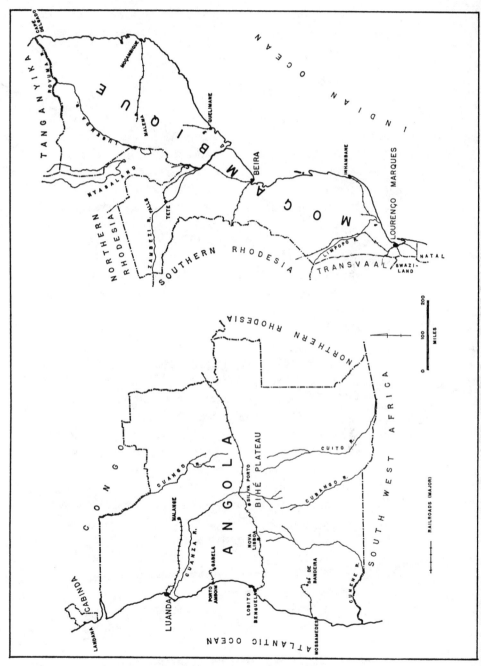

MAP 9   *Portuguese Provinces in Southern Africa.*

gola, on the other hand, has about 5 million inhabitants, and such great densities as those of southeastern Moçambique do not here occur.

| Distances and Prospects

There are some very strong contrasts between the two Portuguese territories in south central Africa. Moçcambique is largely a low-lying, undiversified coastal plain, particularly in the south. North of the Zambezi River, which meanders across this low plain and forms a large delta upon reaching the Indian Ocean, the topography is more diversified. In the vicinity of Lake Nyasa is an attractive hilly land, and the upper part of the Zambezi Valley itself rises toward plateau elevation. The climate in most of Moçambique is uniformly hot and, particularly along the coast, humid, although interior regions have a low annual precipitation average. Only in the higher parts of the northwest is there really abundant rain. Offshore the warm Moçambique Current carries ocean water from tropical latitudes along the low-lying coast. Most of Moçambique is under bush and savanna, and soils over most of the area are poor. Some of the river lowlands are under irrigation, and there are some small intensively cultivated areas immediately behind Lourenço Marques and along the Zambezi, as well as north in the hinterlands of Quelimane and the town of Moçambique.

One of Moçambique's peculiarities is its shape. In the south, the boundary follows along the Great Escarpment as far as the Zambezi River. North of the river, however, Moçambique extends into the higher regions which form part of Northern Rhodesia and Nyasaland, while Nyasaland extends in a peculiar tongue southward onto the lowland to within a few miles of the Zambezi itself. In the north, Lake Nyasa forms the western boundary and the Rovuma River the northern border with Tanganyika. Thus, the territory tapers southward, the northern border being 400 miles in length and the southern (with Natal) less than

50. The Nyasaland border has given rise to the greatest problems, since it has been delimited with the least reference to the ethnic and physical unity of the area through which it lies. For many years, this border was freely crossed by Africans, but in the past decade, the Portuguese have attempted to stop this migration—several times with force. Of all the territories in Southern and south Central Africa, Moçambique's area is perhaps more burdened than any other by boundary liabilities, lack of cohesion, and extensive, remote areas not connected by adequate transportation lines to the main centers. Its great length and small width have much to do with this, as does also the absence of a real core area. South West Africa has a central plateau, on which the capital is located. Even extensive Northern Rhodesia has its economic hub, the Copperbelt, and fairly large capital connected by rail. But Moçambique's capital lies in the very south, where the country is only a few dozen miles in width. Lourenço Marques is more than 1,200 miles from the northeastern part of the country—farther than *any* part of the Congo is from Leopoldville and probably farther than any part of any country in Africa is from its capital. In fact, Lourenço Marques is as far from the mouth of the Rovuma as Nairobi is from Salisbury and farther than Johannesburg, South Africa, is from Elisabethville, Katanga.

Though larger, Angola is a more compact territory. It is nearly square, and boundaries are straight for much of their total length. Unlike Moçambique, Angola is a plateau, much resembling the highveld of the Transvaal and Southern Rhodesia. Around Nova Lisboa the plateau is at its highest, reaching nearly 8,000 feet atop several high mountains but averaging between 4,500 and 6,000 feet over a more extensive region. From this area, the plateau falls in all directions, to the south into Ovamboland in South West Africa, to the north into the Congo basin, to the coastal belt in the west, and slightly to the Katanga in the east, where, however, the elevation again increases. The coastal belt is narrow, broadening slightly behind the capital, Luanda. Largely because of the cold offshore Benguela Current, the coastal plain is dry and desert-like, particularly in the south, where it is an extension of South West Africa's Namib Desert. Being the source area for several great rivers, including the Zambezi and the Kasai, Angola's plateau has a divide character and is not itself blessed with large and fertile valleys. Nevertheless, soils in several parts are quite good, rainfall is adequate for farming in extensive regions, and generally Angola is better endowed in this respect than Moçambique. There are large farming areas in the hinterland of Luanda and on the central plateau. In the north, where the land falls into the Congo Basin, there are dense forests, but southward there is savanna and bush vegetation, with typical highveld grasses on the plateau.

Neither Angola nor Moçambique are very wealthy in terms of mineral resources. There are still possibilities for further exploration and discovery, but there is no Katanga-Copperbelt or Witwatersrand in either territory. Some oil has been discovered in Angola, and there is coal in Moçambique, and in addition a number of minor minerals are known to exist. Notwithstanding this paucity in mineral wealth, both territories have benefited from the mineral riches of the interior plateau. Southern Africa is rather poor in good natural harbors, and the Portuguese possess not only the bulk of the Southern African coastline but also several of the best natural harbors. The shortest distance from the Witwatersrand and Johannesburg to a port is the less than four hundred miles to Lourenço Marques, and Salisbury and the Southern Rhodesian mines lie opposite the port of Beira. The Copperbelt and Katanga lie in the hinterland of Lobito, the port on the coast of Angola. Lourenço Marques, Beira, and Lobito handle a very large amount of bulk material, bringing in millions of dollars annually. Connections to the interior and the ports themselves are constantly being improved. Lourenço Marques competes for the Witwatersrand trade with Durban and for the Southern Rhodesian trade with Beira. The connection with the Witwatersrand is a very old one, having been completed in 1895, but the railroad leading directly from Southern Rhodesia to Lourenço Marques was constructed in the early 1950's. Beira was linked by rail to Salisbury in 1899, and until the railroad to the Copperbelt was linked with the railway through the Congo and Angola to Lobito, Beira remained the only direct outlet for the Rhodesias, although there was of course the laborious journey to Cape Town. After the Lobito link was established, Beira and Lobito shared the exports from the Rhodesia interior, but Beira again attained added importance when the Congo crisis occurred. Although the Katanga section of the Lobito railroad has remained usable throughout the Congo crisis, the possibility of interruption of this service has now focused renewed attention upon Beira and Lourenço Marques.

The cities, Lourenço Marques and Beira in Moçambique and Luanda in Angola, are the focal points of life in these territories. They are prospering because of the economic progress on the plateau. All make a modern, up-to-date appearance, and tall skyscrapers rise in the central business districts. They are the home of the bulk of the white populations in the countries, the terminal points of the railroads from the interior (although Luanda does not lie at the end of the Lobito Railroad, it has its own link with its hinterland), and the administrative centers for the provinces. Angola's Luanda today has over 200,000 people, of whom perhaps one quarter are white. The capital of Moçambique now has a population of about 75,000 Africans and other non-white racial groups, and 25,000 whites. Moçambique's second town, Beira, has over 20,000 people. Moçambique and Angola, which constituted severe economic liabilities to the Portuguese Empire, now constitute one of its economic mainstays, and in recent years to an increasing degree. Immigration has been encouraged, and many Portuguese have since 1945 entered both territories. Many more have gone to Angola than to Moçambique, and Angola in many ways remains more closely tied to the homeland than Moçambique. The white population of Angola is now nearly 200,-000, while that of Moçambique is still under 100,000. The more rapid immigration into Angola has been possible because of the better farmland there, and extensive settlement schemes were devised to accommodate the new arrivals. Actually, the white farmers in the hinterlands of Luanda and Lobito along the railroad did not always meet with success, and many have drifted back to the cities. In Luanda, restaurant waiters are mainly white, often people who have left their failing farm and are now working for a low wage in the city. In Moçambique's towns, the restaurants have African waiters, and behind this apparently incidental difference lie significant causes. The recent strife in the interior of Angola has driven more white people to Luanda, where Africans now experience competition even for unskilled jobs.

The towns of the Portuguese African territories display a wealth and opulence that forms a striking contrast to the backward character of the immediate environs. Lourenço Marques forms a typical example, having recently undergone a period of rapid growth. Both Lourenço Marques and Beira benefit considerably from the tourist trade, for thousands of visitors from South Africa and the Federation travel to the coast for their vacations each July. Lourenço Marques even after 1910 was slow in developing, but from then on, improvements boosted the port considerably in its competition with Durban for the Witwatersrand's trade. The town became the capital of Moçambique in 1907. Although it had been connected by rail to the Witwatersrand since 1895, all cargo had to be discharged by tugs and lighters until 1903, when a wooden wharf was built. This improvement was the beginning of a new era for the place, but the need for further additions to the harbor was soon recognized. In 1914, work was started on the present quay, which is constructed of reinforced concrete and can accommodate 15 ocean-going vessels. There is much space available for further expansion, some of which is presently being carried on. The city lies on the Bay of Lourenço Marques, often called Delagoa Bay, which has been described as one of the world's best natural harbors. Actually, there are some problems, including a narrow channel for navigation through the Bay and a sandbar which is building up across its mouth. The Bay itself is enormous. The island of Inhaca, which is practically connected to the mainland in the south and forms the seaward barrier to the ocean, is about 20 miles from the harbor, enclosing a vast area of water. The Bay is rarely really rough, and any number of ships can seek shelter there. During the war, dozens of ships were in the Bay at the same time.

With the completion of the modern quay at Lourenço Marques and the placing there of several dozen large cranes and a special coal-loading machine, the city's urban expansion went vigorously ahead. There was much swampland even within the municipal boundaries as well as immediately outside the city, and this was drained, reducing greatly the danger of malaria in the area. Although the actual harbor site was a very favorable one, the city itself was in a physically less ideal position, and as in Durban, Port Elizabeth, and Cape Town, land had to be reclaimed. In Lourenço Marques, the land thus added was never put to any permanent urban use, but it is hoped that the expected expansion of the city can be channeled in an organized fashion into this area, which lies immediately adjacent to the present central business district. Meanwhile, a city plan was developed of which many a modern city would be proud. The Avenida da Republica, Lourenço Marques' main artery, is sufficiently wide to permit sidewalk restaurants on either side without crowding the pedestrians, diagonal parking on both sides and parallel parking on the center strip, along with four lanes of traffic. The city plan, in the days of its construction, was considered outrageous, but today it is one of the city's greatest assets. Already, some urban growth had taken place before the modern plan was implemented, and the results of that growth are today visible in the so-called "old town," the area immediately adjacent to the harbor, where narrow alleys and one-way streets give rise to severe congestion. It was well that this sort of urban development was replaced by the master plan now in effect, but even that scheme lacked in foresight. At the time of its construction, no one visualized that Lourenço Marques could possibly grow to its present proportions, and therefore the layout of wide avenues and large square city blocks was terminated by a circular drive, a roadway beyond which land could be occupied by Africans and where they could construct their huts. Today, Lourenço Marques has grown far—indeed, miles—beyond this circular drive, but the lack of planned development has given rise to a haphazard arrangement of residential areas of high and low quality, commercial enterprises, and other mutually detrimental establishments.

This termination of the city plan and the nature of the development beyond it are a source of grave concern for the city-planners of today, who seek to achieve some organization in the outskirts. Already, a city plan for the next half-century has been completed, including a reorganization of the outlying areas. Another effect of the problem has been the development of multistory buildings within the area covered by the initial city plan, the result of an understandable reluctance to invest beyond the organized city's limits. Those structures which have been built are imaginative and attractive, if not always entirely functional, and Lourenço Marques makes the impression of being a much larger city than it really is. The City Hall overlooks the central business district and faces the major thoroughfare leading to the harbor. Adjacent to the City Hall are the unusual cathedral, the Radio Clube de Moçambique, and the botanic gardens. The Radio Clube has a powerful shortwave station on which is broadcast a commercial program heard all over Southern Africa. In addition, there is a local station broadcasting in the Portuguese language. The Radio Clube runs a symphony orchestra which has a weekly concert open to the public and a season that is no less than 47 weeks in length. The building of the Radio Clube is one of the most imposing in town, rising above the botanic gardens, culminating in a tall tower.

For a city of its limited size, the central business district is very extensive. There are three distinct sections here. One is the "old town," which lies next to the harbor and includes some of the oldest buildings in the city. To the northeast of the "old town" is the high-class part of the district, where most of the tall buildings are, the high-quality stores, department stores, large sidewalk restaurants, wide avenues, and where the buyers are mainly white people. In the northwest is the section of the central business district where stores are owned by Pakistani, Indians, and Chinese, and where non-whites do most of their shopping. Approximately in the center of the whole district, where the three divisions come together, is the market. Here, farmers from the environs of the city sell their products, fishermen their day's catch, and Africans their handicraft articles. Everyone buys at the market, and it is here that white and non-white really mingle. The market from dawn to dusk is a hub of bustle and activity, crowds pushing each other amid the sometimes unbearable stench of the fish and the squeals of live pigs.

The Portuguese advertise Lourenço Marques as a "continental city in tropical Africa," and in some respects the description fits. The outskirts of the city are studded with African huts made of thatch and reed and often gaily decorated: there is nothing "continental" about these parts of the city. Neither is there anything un-African about the market. But the sidewalk restaurants on the Avenida da Republica (the Continental and Bambu) and the ornate buildings, the noisy, crowded sidewalks, the splendid hotels overlooking the Riviera-like beaches and the Bay and the palm-lined boulevards—all these features are most reminiscent of the best Europe has to offer. Lourenço Marques must rank among the world's most beautiful places. Along the wide avenues are magnificent flamboyants which color the town a bright red at blooming time. Much of the city lies on a high sandbar, the bluff leading down to the water's edge being steep and bushy and over much of its length terraced into spectacular gardens. Tall buildings dominate the edge of this terrace, and here are the hotels famed throughout Africa: the Polana, Cardoso, Girassol, and others. From several of these, the harbor and central business district can be seen, ships lying in the Bay awaiting docking berths, and the new, sparkling white or brightly colored, dozen-stories-high buildings along the Avenida da Republica dominating the scene. Lourenço Marques has character. South African cities of similar size so often appear barren of vegetation, dusty, unkempt, and featureless. Lourenço Marques is full of variety and attractions. There are the bull ring, the old fort, excellent swimming pools, picturesque architecture, and much more than would be expected in a city of just 100,000 people.

One man has placed his stamp upon the whole city of Lourenço Marques in unmistakable fashion. He is the famed architect Miranda d'Alpoim Guedes, known to his fellow citizens mainly as "Pancho." His controversial and vastly interesting styles have received world-wide attention, and in a way he has used Lourenço Marques as a laboratory within which to experiment with his ideas. Whether or not his brightly colored buildings with apparently outrageous decorations are appreciated, it cannot be denied that they form a most prominent part of the city's aspect. Most of his more individual creations defy description. Apartment buildings are adorned with a series of spines protruding from walls painted in a bright multicolored mosaic. Homes are given chimneys shaped like tall mushrooms. His structures are a real asset to the city, since they greatly add to the variety of the townscape. Largely as a result of his labors, parts of Lourenço Marques are probably years ahead of their time in terms of architectural appearance, something which, here in this remote part of Africa, is not expected.

It is not only Lourenço Marques, however, which makes a progressive appearance among the towns and cities of

Portuguese Africa. Beira, with only about a quarter of the population of Lourenço Marques, also has its impressive buildings, attractive central square, and sidewalk restaurant. Beira and Lourenço Marques, competitors for the trade from the interior, have in common the hot climate of the coastal regions. Day after day, the sky in these regions is blue, the sun beats down and raises the temperature to the 90's, and the slight seabreeze, bringing only mild relief, carries moisture into this heat, making conditions even more uncomfortable. There are some who thrive in this kind of weather, but most white people suffer, particularly during the summer months of December, January, and February. Generally, the days are merely hot, in the 90's and perhaps close to the 100-degree mark, but on occasion a real heatwave sends the mercury up as far as 115 degrees, or, as on a day in September (spring), 1958, to over 121 degrees in the shade. These extremes may be rare, but hot and humid days are common. Luanda, in Angola, is somewhat more comfortable, although it lies fully 10 degrees of latitude closer to the equator. Here, the offshore cool waters of the Benguela Current help to keep not only the temperature down but also the relative humidity, a vital factor in the environmental climatology of the tropics. Luanda, also, is a much bigger city than either Lourenço Marques or Beira, but it has much in common with these places. Like its rivals on the Indian Ocean, Luanda makes an unexpectedly prosperous appearance, with its magnificent shore drive, large apartment buildings, and busy city center. At Luanda are some sandspits building up into the ocean, and they are used for the construction of quays. Luanda is not blessed by natural harbor conditions as is Lourenço Marques.

Unlike so many of the bleak, desolate small towns on the highveld, even the smaller villages in Moçambique and Angola are picturesque, neatly laid out, lined with trees, and possess some real individualism. Here, there are not the corrugated iron shacks, characterless houses, and the windswept veld all around. The Portuguese have a remarkable

364 | AFRICA SOUTH

ability to make something out of very little, and to give
something unusual to even the most insignificant settlement.

| Assimilation and Forced Labor

The composition of Portuguese Africa's population is difficult to assess because the Portuguese have a history of
non-racialism and intermarriage, resulting in an absence
of statistics on racial groups and a sizable mixed population. The statement that the Portuguese practice non-
racialism does not imply that they do not discriminate
against Africans. Individual Portuguese are indeed guilty
of such activities, but there is not, as in South Africa and
Southern Rhodesia, a strict separation and classification of
the racial groups in the country. On paper, the Portuguese
intentions with reference to the African peoples under their
jurisdiction are reasonable enough, when compared to British and French policies over the years. In actual fact, however, the African in Moçambique and Angola is probably
worse off than his South African or Southern Rhodesian
contemporary. A very few have achieved the virtually impossible, this being to gain a higher education as well as
attain full citizenship. These few are always held up as examples of what Africans can attain if they only make the
effort, and the system of administration is then judged by
these individuals. However, the odds against an African
making educational and economic progress in Overseas
Portugal—particularly in Moçambique—are tremendous.

Portuguese colonial policy is famous for its "Assimilado" system. Article III of the Political Constitution of
the Republic of Portugal determined that all Portuguese
citizens, whatever their race, shall have equal rights. A subject of Portugal is not necessarily a citizen of Portugal; an
African must fulfill certain requirements to become a citizen. To attain this goal, an African must speak Portuguese, he must live in a white man's way, which is taken to
mean such things as sleeping in a bed with sheets and blankets, eating with a knife and fork, and dressing in a West-

ern manner, and in addition he must practice a recognized profession or possess property which allows him to maintain himself in a "civilized" manner. If able to fulfill these qualifications, an African may apply for assimilado status. If he succeeds in getting it (there are less than 8,000 assimilados among Moçambique's 6 million Africans), he is generally accepted as a member of the Portuguese community and is accorded most of the rights of the white Portuguese. He may purchase a home or reside in a white area, and all but some government jobs are open to him. The assimilado is evidence that the Portuguese have no such extreme racial phobia as do the Afrikaners and that they are willing to accept Africans as partners in life— equal partners on the surface if not in their own minds. Recently, with the influx of large numbers of peasants from Portugal, race antagonism has shown its head, particularly in Angola. It is also a matter for contemplation what effect a much larger number of assimilados would have had upon the Portuguese citizens who are willing to accept a few. In actual fact, only one or two houses and apartments in "white" Lourenço Marques are African-owned, and the number of whites in the town is several times as large as the number of assimilados. But the system exists, and Portuguese point to it in defense of their treatment of the African masses. The philosophy is that those Africans which are not assimilados have not shown the interest and are unwilling to expend the effort to attain that standing. In reality, the obstacles before the African in his quest for full citizenship are stupendous. It is, also, significant that a number of Africans who could become assimilados do not make use of the opportunity. There are several of these in Lourenço Marques, for instance, who say that although they have all the required qualifications, they are Africans and could never become Portuguese. They are willing, in the interest of their growing nationalist philosophies, to give up all the privileges which assimilado status entails. They say that the future is theirs and that when it arrives, they would like to see their past uncontaminated.

In Portuguese Africa, and perhaps in the whole Portuguese domain, this is something new. The assimilado system has long acted as a safety valve, has given the African something to aspire to, and the assimilados themselves have been living proof of the wisdom of Portuguese policy. In the schools of Luanda, white and black can be seen to play together, though not in the ratio of the population's racial divisions. True, a few Africans are being taken up in the Portuguese way of life, but it must not be forgotten that the four or five African schoolchildren in a class of over a hundred represent millions who have never seen a school—or, for that matter, a hospital. The assimilado system is one small part of Portuguese policy in their African territories. By virtue of its contrast with what has happened in several other parts of Africa, it is outstanding, on paper it is spectacular, and it is often quoted as evidence for the tranquility which has so long reigned in Moçambique and Angola. Portuguese say that Africans see their brothers as assimilados, as Portuguese citizens. Therefore, any racial friction would become a matter not between black and white but between civilized and backward, having nothing to do with race. And since there are many countries where very rich and very poor, very advanced and very backward, people live side by side in harmony, it should not surprise anyone that there has been no trouble in Angola and Moçambique. The Portuguese feel that they have put the onus on the Africans. They have nothing to complain about, they say, because the opportunities are there for the taking.

No doubt a great majority of the Portuguese whites are in fact opposed to racism and practice what they preach in everyday life—to a point. In the cities, some of Africa's apparently most hopeful scenes are enacted each day as waiters serve white and African alike in the restaurants and as schoolchildren work and play together, whatever their skin color. Again, it is the difference between the city (which is by no means devoid of its problems) and the hinterland, the interior, which is so great as to be virtually incomprehensible. Interior Moçambique and Angola seem for centuries to have stood still. Portuguese administrators govern with an iron hand the isolated districts where Africans have barely emerged from the way of life of hundreds of years ago. Subsistence agriculture is the rule, education remains virtually absent, mission stations are few and far between, and they, anyway, do not contribute much to the African's progress; they are simply unable to do so where the will does exist. Health conditions are as bad here as anywhere in contemporary Africa, and hospitals are likewise most uncommon. The administrator holds absolute power in his district. He sentences and punishes offenders, collects taxes, regulates all trade, assigns cropping areas, and decides what crops shall be grown. He also issues permits for travel, no inhabitant of his district being allowed to leave or re-enter without his permission. Perhaps most important of all, he decides whether an African has achieved the status of assimilado, for he rules on the application entered for this purpose.

The administrators are feared by all their subjects, but it would not be correct to assert that they are all invariably tyrannical despots. For some, this description would be excessively mild, but there are a number who are respected and admired by their wards and who have the interests of their subjects at heart. The responsibilities of the administrators are many and varied, and it is not surprising that relatively few men have been able to rise to the occasion of their appointment and adequate performance of duty. Those who have are respected and admired not only by the Africans in their district but also even by those Portuguese who are critical of the treatment of Africans in the two territories. A successful administrator can place his stamp on the whole area under his power. His knowledge of the land and crops can avert famine. His fairness in dealing with African disputes can win the respect of the villagers. His treatment of offenders can gain the approval or reap the fearful hatred of his wards. Frequently, a successful administrator attributes his good results to a philosophy which is very common among the Portuguese. All indigenous Africans are "children," they feel, and should be treated as such. They assert that the Africans are likely to remain children for many decades to come, and so the Portuguese have come to ingrain their actions with this philosphy of paternalism, sometimes the most ruthless paternalism that can be envisioned.

Good and bad administrators alike contribute to the desperate unhappiness of the Africans in the interior. On the surface, the local inhabitants may appear cheerful enough, saluting the passing car with a broad smile, removing their hats and waving long after it has gone past. But here in the interior, and particularly in Moçambique, survives a form of forced labor which has been referred to by several observers as neo-slavery and which is strongly reminiscent of the horrors of another century. The labor code enforced in the Portuguese colonies was issued in 1899. This code states that Africans are obliged to attain, through work, improvement in their social condition and that "they have full liberty to select the mode of fulfilling this obligation, but if they do not fulfill it by some means, the public authorities can impose its fulfillment upon them." It is, of course, the administrator who in his district decides whether people are working or not. If labor is required, a number of Africans can be declared idle, and they have no recourse to a court of law to defend themselves against the allegation. Upon being declared idle, the African is given the choice of signing a contract for a private firm, which involves his going to work at a minimal wage until his services are no longer required or until six months have elapsed. If he does not sign the contract with the private firm, the African is made to work on such public projects as a road, dam, or bridge. His work is declared to be in the national interest, and it is felt that he has to do his share for the country. His wages are pitiful, less than ten dollars per month with in addition some clothing and meals of porridge and bread. Conditions in the country have been compared to those under which chain gangs toil, and those who labor in the cities may be made to work twelve-hour days on the docks, on street-construction, on trench-digging, and on various other hard jobs of this kind. Several aspects of the system are particularly unsavory. Even though the system has been described in its worst form by several authors who see only one kind of Overseas Portugal, it is not universally accompanied by violence, raids on villages, beatings and terrible hardship. It often is, but this is the result of the manner of action of the individuals who implement the law. The worst feature of the labor code is the utter inability of the African to defend himself against the charge that he is idle. The fact is that the system reduces the Africans in the territories to a serfdom from which they cannot extract themselves, and their own chiefs participate with the administrators in rounding up the so-called idlers. In order not to be considered an idler, an African must meet one of a number of conditions. Even though such fulfillment does not guarantee immunity from the system, it helps to avoid being declared idle. These conditions include being self-employed in a recognized economic activity, having worked as a laborer within the past six-month period, possessing 50 head of cattle or more and herding them, having worked in the past six months in South Africa or Rhodesia, or having within the past year completed military service. In addition, an African may avoid being classified as idle if he can prove himself to be what is locally called an "African Agriculturalist," which means that he must plant cash crops and use farm machinery. This alone is not sufficient; he must appeal to the administrator of his district for the "Agriculturalist" certificate. If the administrator has a lack of idle men in his district, it may require him some years to decide upon the application—within which period the applicant is liable to recruitment as a private or state laborer.

There are only two ways by which an African living in his rural district in his own village can hope to fortify himself against the feared classification as idler. One is by becoming an "Agriculturalist," the other is by raising at least 50 head of cattle. The cattle population, as Marvin Harris points out in his Report on Labor and Education in Moçambique, is such that if all the cattle owned by Africans were divided into herds exactly 50 head strong, a mere 12,-000 to 15,000 Africans would possess the number required by the authorities to avoid idleness. Of course, the African-owned cattle in the territory are not divided thus, most African villagers owning just a few, and some a number much larger than fifty. Hence a mere handful of Africans can satisfy the requirements of being a cattle-raiser, and since the number that have attained "Agriculturalist" status is likewise small, the great majority of the Africans find themselves, if they attempt to live permanently in the village and district of their birth, susceptible to recruitment.

The other way by which an African avoids this kind of forced labor is by volunteering for other kinds of labor before he can be recruited. The entire system is one aimed at keeping the African population a conveniently subservient labor reserve, and as such it has been a great success. Africans can join the army, they can sign a contract to work in Johannesburg, on the Witwatersrand, or on the Great Dyke, or they can attempt to find work in Moçambique itself as domestic servants, clerks for the government, messengers, and such. It must always be remembered that government and white population can create only a small fraction of the jobs required to engage the entire labor force: there are nearly six million Africans in Moçambique and only some tens of thousands of whites. There is thus little chance of escape for the villager. Unless exceptional circumstances allow him to break through the barrier of serfdom, his position as a legally unprotected tool of the state is complete. The intricacy of the system by which this arrangement is maintained is considerable, and countless avenues of apparent hope for escape are meticulously closed by administrative manipulation.

The system, through the years, has brought Moçambique and Angola much revenue. Causing the Africans to seek employment for fear of being recruited at slave wages and preventing them from living a life of agricultural subsistence without the dangers of such conscription always hanging over their heads, it has resulted in a migratory labor force of great proportions, always available when wanted and always profitable when utilized. Africans who work in South Africa or Rhodesia pay taxes, and the Portuguese are paid a sum of money for every laborer who crosses the border. There are nearly half a million such workers on the plateau, and they have brought the Portuguese government millions of dollars, while returning with their own wages to be spent in the territories. There are agreements between South Africa, the Federation, and the Portuguese government concerning these labor arrangements, and since men aged eighteen to fifty-five are susceptible to being declared idle, there is never any trouble finding migrants. Moçambique has twice the African population of Nyasaland, and it does not have the cash cropping that at least some thousands of Africans carry on in the latter territory; hence it forms the greatest labor source on the subcontinent. Not only do the laborers sign contracts at recruiting stations of the Transvaal mines and other enterprises scattered throughout the territory south of the Zambezi, but a number cross the South African border illegally and seek work on their own, thereby avoiding the possibility of being declared idle and being put to work in Moçambique. Thus the result is a stream of migratory labor with favorable consequences for the Portuguese and the white South Africans. In the districts and villages, the effects are disastrous. It is claimed that in some districts three out of every four able-bodied men are absent at any given time, and even if the figure averages 50 per cent, the results in terms of village and family life, as well as progressive farming, are devastating. While the man is gone, the woman of the house is left to do the farmwork, and she has no incentive to rise beyond the subsistence practices which drove her husband and sons away. Hence no real progress has been made in the field of farming in the interior, a factor which in the future may prove to have se-

vere negative consequences. If the system of migratory labor should for some reason suddenly come to a stop, and
should the labor force return to the interior in its entirety,
the land will have to carry more people than it has done
for many decades. During those decades there has been
virtually no progress in farming, soil conservation, cropping practices, and the like; the results may be calamitous.

Portuguese officials state the case for the system as elo-
quently as defenders of Apartheid defend their standpoint.
Certainly, the African has not over-exerted himself to secure his own progress, and he has not voluntarily toiled in
the fields to wrest the best yields out of his land. Once again
it remains a matter of opinion whether there are moral
justifications for denying a man to be idle, when through
his idleness he becomes a liability to the district and the
state. The Portuguese say that they, themselves, have
achieved what they have through hard work, and why
should Africans not be made to do the same? There is no
excuse for idleness, and children should not be allowed to
remain idle. Africans who are not assimilados are children,
and they should be treated as such. It is an astounding commentary on the Portuguese mentality that those who can
accept assimilados as neighbors and fellow citizens should
see nothing wrong in a system which does more than anything else to prevent the rise of the African to better ways.
In that part of Moçambique which lies north of the Zambezi River, a system of compulsory cropping (of cotton)
is maintained where a migrant labor force is not required.
Here the Africans are not forced to leave their farms, but
the progress of farming has received a blow similar to that
in the south. So much acreage has to be placed under cotton
that the food acreage is drastically reduced; when the cotton is sold to local mills, there is no way in which the African farmer can seek higher prices by going to a competitor,
since all are government-owned. As a result, the acreage
under food crops was dangerously reduced, in some places
disastrously reduced, with famine the result. Africans have
no insurance against crop failure. They have the alterna-

tives of receiving minimal payment for their harvest or having crop failure and no funds with which to buy the food they were unable to grow. The government has none of the risks; the African risks his health. There have been outcries, even within Moçambique itself, against the system, and Africans have sabotaged it by refusing to plant the seed, failing to attend to the crop, and by carefully boiling all seed provided before demonstratively planting it. However, the Portuguese have their measures against this kind of resistance. Corporal punishment which is not soon forgotten is imposed upon a delinquent farmer, with the result that most have been kept at work. The scheme has recently been expanded also to include areas in the southern half of Moçambique, where practices have been similiar. It is a known fact throughout the territory that Africans are not given a fair deal by those who weigh the crops they bring in and that they are cheated without any opportunity for legal redress. African farmers have trouble making their land yield sufficiently to weather the dry season; the wholesale planting of much of each individual's land to cotton and the expenditure of energy in raising the crop means misery to many. There are indications that things have slightly improved, that prices paid for the crops are now somewhat higher, and that some Africans are able to buy more food than they might have been able to grow on their land. However, the system, dependent as it is upon the fairness of the officials running it in each district, and controlling as it does the African's activities on his own land, is never likely to rank among the moral triumphs of colonial Africa.

Although Africans in such agricultural activities as cotton-growing and in employment in the cities are relatively safe from enlistment in the works gangs of the state, even they can be forced to join. The domestic servant in the city provides an outstanding example of the intricate system of controls the Portuguese have established to remain capable of enforcing their every will even upon wage-earners. Africans who wish to become domestic servants in the employment of white families must get a work permit upon which such matters as his salary, the nature of his work, and the address of his employer are all recorded. An African of, say, age sixteen who finds an employer willing to hire him must go with his employer to the authorities, where he and his future boss must sign a contract. For years such a young servant—known as a *piccanin*—could start with a monthly salary as low as 75 escudos per month, or about $2.50, plus room and board. If he and his employer agree that the contract will be renewed the next year, they again must go together to the authorities and renew it with their signatures. The contract is favorable to the employer, who may fire his servant without notice at any time, while the African employee may not leave except at the end of the contract period, when he can refuse to sign, taking the consequences. There is an avenue via which the employer can complain about the poor quality of his servant and have him punished, while the African on paper has the right to complain about cruelty or any other mismanagement on the part of his employer. Needless to say, not many Africans lodge complaints against their masters to whom they are bound by contract. If they do, or, more likely, if an employer complains, the police investigate the matter. Should the African be found at fault, he is taken to the police station and is there flogged with an instrument called a *palmatorio,* a sort of wooden bat with a handle, the flat part of the bat being perforated by holes. Blows of this instrument are exceedingly painful and leave large welts which, untreated, develop into sores. The African, hence, does not forget his offense easily. The employer, conveniently, has not had to impose the beating himself. Actually, there is a rule that prevents employers from privately beating their servants, but it is not strictly enforced.

When the *piccanin* turns eighteen, his salary may still be as little as 200 escudos per month (under $7). Whatever his salary, however, he must now pay an annual tax of 300 escudos. Even if he is without employment, he requires the 300 escudos to avoid imprisonment or, of course, inclusion in the labor gangs. Africans without jobs in the city are removed in this manner as idlers as fast as those who live in the interior, and even if somehow an African without employment possessed his tax money, he would soon be taken away. Those who are fortunate enough to retain a permanent job in someone's employment have their salary slowly increased by annual increments, under normal circumstances. An increase in salary of 25 escudos (about 80 cents) per month each year would be considered good. Actually, in recent months there has been a change in the minimum salaries for *piccanins* and also a tendency to pay servants somewhat more at a faster rate, but still, increases come in dimes, not dollars. A good salary in Lourenço Marques for an African domestic servant is 500 escudos (under $17) monthly, plus room and board. Although there does not appear to be a legal ceiling to the salary earned by servants, there is in practice a limit, maintained by the authorities before whom employer and servant sign their contract. It is not permitted to pay an African more each month than his contract indicates, so that the opportunity for the servant's material progress is rigidly limited.

The recent elevation of lowest salaries for *piccanins* and increase in salaries for longtime servants have had negative effects as far as the Africans are concerned. Instead of raising the salaries of all, the result was that many were fired and a few remained to enjoy the new privileges, such as they are. The point is that many low-income white families were able to afford two or three *piccanins* while they could be hired for as little as 75 escudos per month. When the minimum salary was increased, they decided they could afford only one, and the rest were sent searching for other employment. Suddenly, then, a large number of Africans were unemployed as a result of pay raises. This was something entirely new, since for many years there had been a shortage of domestic servants. Needless to say, many of those turned out as a result of the new measures were before long drawn into the labor gangs. If the unemployed were eighteen or over, they had to pay the tax, and without

376 | AFRICA SOUTH

work they could not. Thus they came to owe the state
money, which could be earned by the compulsory work on
the gangs. These, however, pay very little, and it takes
many months before an African can save his tax money.
Having thus collected his arrears, he is probably in debt
for the next year and must continue to work in the gangs.
It is difficult to be sure whether the sudden new policy of
higher pay was in fact meant to improve conditions for African domestic servants or whether it was aimed at the enlargement of the labor gangs, because perhaps of a lack of
idlers from the interior. Whatever the intent, the results
are clear. What is astounding is that such a system has survived as long as it has, with so little notoriety, while the
world concentrates its attention on the plight of the African in South Africa, who is immeasurably better off.

| Only Provinces

In spite of his state of serfdom, in spite of all the cruelty
and the hardship imposed on so many Africans, one can
gain the impression that race relations in Portuguese Africa are not bad. Many people refer to the almost tangible
difference between the conditions they leave behind in South
Africa and the apparently much more pleasant atmosphere
in Moçambique. It is astounding how Africans from Moçambique at work in South Africa, upon seeing a car with
Portuguese license plates, come broadly smiling, to con-
verse with the people from "back home." Why is there not
the hatred and resentment in every African, few of whom
can have escaped all impact of the system under which they
were born? What is it about Portuguese Africa that sub-
merges the harshness of the system, leaving an image of
friendliness, interracial cooperation, scrupulous fairness
and honesty? Part of it seems to be the African's philosophi-
cal reaction to the entire situation. Barring a few active
antagonists who, within and without the territories, agitate
against the prevalent conditions, it would seem as though
the Africans have simply given up the hope of betterment

and take everything that happens to them as a natural event, to be expected just as summer rains, the hot sun, and death. They seem therefore able to muster up an amazing cheerfulness with which they face all that befalls them, and they have learned not to murmur, a lesson of centuries. They are herded as easily as sheep, and with them the attitude of resentment, reproach, and bitterness over their lot is largely submerged if not absent.

There are, of course, exceptions to this state of affairs. In northern Angola the effects of an independent African state adjacent to Portuguese Africa are being felt as Congolese Africans support Angolan Africans' insurrection. Conditions in Portuguese Africa appear to be better known outside Angola's and Moçambique's boundaries than within, and with Moçambique's newly independent African neighbor, Tanganyika, trouble is to be expected. But the bulk of this trouble is made from outside the territories. Even those Africans within Moçambique and Angola who would oppose Portuguese policies actively are rendered harmless by colonial Africa's most complete control, and African political organization is here unknown. Agitators have appeared, but they are removed from the scene and often never heard of again. Portugal's colonial administrators have no patience whatever with African nationalism, and they do not hesitate to eliminate it. The brutality committed by both sides in the action in Angola evidences the Portuguese attitude as well as the terrible force of eruptive hatred now suddenly released by the Africans. In Moçambique, particularly in Lourenço Marques and its environs, it is hard to believe that the seeds of this hatred also exist. Hard to believe, that is, from the appearance, demeanor, and friendliness of the Africans. When the social condition of these people is observed, it is difficult to understand that it has not produced violent resistance at a much earlier time. Part of it must be the success the Portuguese have had in maintaining the isolation of their territories. Ignorance is bliss, and the African's ignorance of what they are missing in terms of health, education, and material progress is bliss to the Portuguese, who have long benefited by it. But times are different now, and the Wind of Change is blowing hard into both Angola and Moçambique. So far, the geographical fringes are most affected, but throughout Portuguese Africa, change is in the air. The Portuguese themselves admit that change is needed and state their intention to promote and guide it. After wasting their chances for centuries, they are now too late. Even if the Africans within the territories would remain servile for a sufficient length of time to permit the Portuguese to carry out their plans, outside forces will come to interfere. This part of the buffer zone is very vulnerable to the impatience which emanates now from a united Africa to the north. Within the Portuguese territories, the change will be all-encompassing, whatever military strength Portugal deploys in order to resist it. For many Africans, the change will be a tragic one. The Africans of Portuguese Africa, held under control so completely for so long by so few, have become a mass of sheep who are influenced perhaps more easily than Africans anywhere else. Whoever gains the power in these countries will hold sway absolutely over the people, who have not known and do not know popular resistance, cannot express and do not have a common opinion, and do not know the power of combined and unified political action. One of Portugal's legacies may be an African people that is ruled—indeed, may have to be ruled—with a hand which, though probably African, will be stronger even than that of Portugal.

Even within Moçambique and Angola, but more so in Moçambique, there is opposition among the whites against the longtime policies of the Salazar government. Angola has always been tied more closely to the motherland than Moçambique, and it is in Moçambique where the greatest opposition among whites to government practices has developed. Here, elections have actually proved an anti-government majority in certain districts, and it is probably true that here the Salazar government has more antagonists than supporters. Those who are the liberals here want cooperation with the African majority and want to develop the country along Brazilian lines. This, of course, is wishful thinking, Moçambique's known resources being what they are and the African having experienced what he has. Although a number of the assimilados could accept the purpose of the idea and would no doubt cooperate, the future of white-African cooperation in this territory is bleak. In the future it is not merely a number of assimilados who will express the Africans' views. Some Africans already consider the assimilados to be traitors who have advanced by the white man's graces and who have collaborated in the suppression of the masses. It would appear likely that the future in Moçambique belongs not to the Portuguese, neither to the assimilado, but to the extreme nationalist African who will muster support by any methods.

During the past ten years, Portuguese efforts to prevent news leakage from the overseas territories have failed repeatedly. The Portuguese have widely advertised such achievements as the establishment of agricultural villages to be occupied by white and African farmers alike and have attempted, by statistical publications, pamphlets, and other propaganda material, to convey to the outside world the image of a new Overseas Portugal. In the rising tide of criticism, these attempts now appear futile, and the state of affairs in Moçambique and Angola is now being attacked from all sides. In the United Nations, Portugal, a staunch defender of South Africa, has found itself completely isolated, casting a solitary vote against unified opposition. The United Nations has shown an ever greater interest in conditions within non-self-governing territories, and in order to avoid inspection, Portugal's goverment officially changed the status of the territories to "provinces" of the mother country. This occurred in 1951, and throughout the subsequent years of attack upon colonial powers and colonies Portugal has maintained that it possesses no colonies, only provinces which happen to be separated from the motherland but which have the same status as provinces of European Portugal itself. Technically, therefore, to speak of "non-self-governing territories" is a contradiction in terms; according to Portuguese opinion, since Portugal now consists of a number of provinces, it has no such areas under its jurisdiction, and therefore United Nations intentions do not apply in its case. Portugal, like South Africa in the matter of South West Africa, has been unwilling to cooperate with the United Nations in any way as far as its overseas possessions are concerned, and its policies have undergone severe censure. Portugal has refused to permit investigation of the terrible violence in northern Angola, which broke out on March 15, 1961, and has since then claimed an estimated thirty thousand lives. An increasing number of observers have reported upon conditions in the territories, and few of these reports have been favorable. It remains true, nevertheless, that even Portuguese Africa can present a semblance of real harmony, progress, and good will. Portuguese officials will proudly show the visitor the model township for Africans outside the circular drive of Lourenço Marques, the racially mixed farming projects in the hinterland of Luanda (closing an eye to unequal land division helps here), the restaurants where white and non-white are served without discrimination, the assimilados participating in "civilized" life, the dam projects, roads, and thriving business districts in the towns. They insist that whatever the future of the provinces, they are now eternally Portuguese, and so are the African inhibitants. That the number of assimilados remains so small and that this may be because of not only inability but also unwillingness on the part of Africans to join an alien culture is not considered. There are few assimilados because Africans are lazy, according to the Portuguese, and it has nothing to do with lack of opportunity nor with deliberate unwillingness. The Portuguese remain convinced that almost all Africans take pride in their Portuguese heritage and that any insurrection, present or future, can only be the result of outside influence and cannot have anything to do with internal conditions. In Moçambique particularly, a few administrators and one or two churchmen have stated the repulsive nature of prevalent conditions, but even though there may be among the whites, a majority of opponents to what is happening, most cannot afford to speak.

Portuguese Africa is a land of paradoxes. Here are territories called provinces of a European country which display more of the old colonial conditions than any other in 20th-century Africa. Here, where the freedom and rights of the individual are at a minimum and where the African has been kept deliberately uneducated and backward, it is possible to see what in milder South Africa one can never see: children of all races playing together and learning together, mixed marriages without official disapproval, and Africans participating in many of the spheres of life which in other parts of colonial Africa have largely been reserved for whites only. Here in Overseas Portugal it is quite impractical to speak of pure whites, for the majority of Portuguese possess some non-white blood and are not at all concerned over it. A Portuguese resident from Europe is generally classified as white (European), but a large number of persons of mixed race are accepted as Portuguese citizens by birth, as opposed to "assimilated" Colored and Africans. Even among the assimilados there are a number of persons of mixed race, reducing even further the body of true Africans who have achieved this status. The Portuguese have a remarkable ability to make the most of their achievements. For long, the small group of assimilados has helped maintain the image of utter fairness of the Portuguese regime. No one seemed to notice that the number of these people did not grow much. While South Africa was blamed for its African education policies, a few African schoolchildren in white schools produced an image of "education for all in Overseas Portugal." Economic development through foreign investment was long discouraged, to the detriment of the economic evolution of the provinces; yet visitors would report in glowing terms of the material progress as evidenced in the modern cities. The true conditions within Portuguese Africa are now being recognized, and, like the South Africans, the Portuguese have stout defenders and strong opponents. There are times when the degree of disagreement between these two camps is so complete that it seems impossible that they should refer to the same area. The great diversity of conditions within the provinces enables both sides to make claims. Whatever the outcome of the debate, it cannot be denied that the mass of Africans have silently suffered the effects of a system which has deprived them of the most basic human rights. Portugal's African provinces have one or two showcases which illustrate what Portuguese idealism recognizes as progress, but they are rare instances in an ocean of isolation and deprivation.

FURTHER READING

DUFFY, J. *Portuguese Africa*. Harvard University Press: Cambridge, 1959.

———. "The Dual Reality of Portuguese Africa," *The Centennial Review,* Michigan State University, 1960.

HARRIS, M. *Portugal's African Wards*. American Committee on Africa: New York, 1958.

KIMBLE, A. *Tropical Africa*. 2 vols. Twentieth Century Fund: New York, 1960.

The Beloved Country

Southern Africa has long withstood the changes which must inevitably come everywhere on the continent. It has faced the onslaught of the forces of African nationalism with a degree of success that has confounded all those experts who in 1953 predicted that the Federation would not last a year, who after Sharpeville gave South Africa only months, and who early in 1961 saw the violence in northern Angola as the end for Portugal in Africa. Rapid change has been the order of the day in Africa, as evidenced by the early independence of the Congo after the Belgians just months previous were talking in terms of decades, and the formation of the French Community when it appeared that France had as little foresight in Africa as has Portugal. But the Wind of Change has to blow harder and even more persistently than ever if it is to have an equal impact south of the Congo and newly independent Tanganyika. It now meets its greatest obstacle, penetrating the white man's stronghold of centuries. It has already ruffled the buffer zone, but the surface calm of South Africa itself has not been severely disturbed. Here in Southern Africa is concentrated the greatest number of white people in all Africa, and there is greater determination to prolong the status quo than anywhere else. In Kenya, Tanganyika, Chad, and

Congo the whites were always able to return to their homeland in the event of serious political upheaval, and many
have done so. Even the whites in Angola and Moçambique,
and to a certain extent the Central African Federation, can
speak of having a European homeland. Not, however,
those of South Africa, and there are over 3 million whites
in that country. The great majority of these people do not
know another fatherland than their Republic, and for them
there is no avenue of egress.

In the south of Southern Africa, African nationalism
stands face to face with an enemy it does not seriously face
elsewhere on the continent: white nationalism. True, there
are always a few whites in Kenya who proclaim that they
are as African as black Africans since they, too, were born
in that country, and in practically all other African territories there are people with similar sentiments. But once
deprived of their safety, comfort, and privileges, very many
of these whites have abandoned their philosophies and have
left for the country they used to call "home." White nationalism is serious only in South Africa, and it has real foundations only in the Republic. A few English-speaking South
Africans still see Britain as the real homeland, but the
great majority of the white people in the country have only
one allegiance, and that is to the flag of South Africa. It is
in South Africa that all the extremes in the political scene
of the continent reach their culmination. Here, white and
non-white have drifted as far apart as they have in any
region of the world. The solution for the future, if there is
one, will be unique. This is one of the grounds for justification frequently brought up by defenders of the Bantustan
plan: the problem is unique, and the scheme of regional segregation is also. Hence, criticism is unwarranted and com-
parisons to events in the other countries of Africa are ir-
relevant. That the Afrikaner highway to harmony is seen
by many as a road to irreparable racism is thought to result from lack of insight; South Africans really do believe
that white and black nationalism can live side by side—
literally.

Between the extremes of African nationalism in the north and white nationalism in the south lies the region which is at the same time a remnant of colonialism and the scene of successful African opposition to white rule: the buffer zone. Although only Britain and Portugal among the colonial powers are immediately involved in the events occurring in this area, there is much more to the political picture than the effects of British and Portuguese colonial rule. In the Federation of the Rhodesias and Nyasaland, something closely approaching Afrikaner sentiment exists in the south while the power of African nationalism is displayed most prominently in Nyasaland and the north. At times the contrasts between conditions in Southern Rhodesia and the two northern territories appear to be so great that it seems inconceivable they should occur in one and the same country. The Federation, then, possesses the characteristics of the Africa north of its borders as well as those of the south. The Federation differs from independent Africa in that the Africans have not been granted total independence and that the whites retain much of the power in the north and most of it in the south. But even in the south, the differences between the Federation and the Republic of South Africa are obvious, notwithstanding the existence of Afrikaner-type philosophies. A majority of white Southern Rhodesians are seeking a formula which will admittedly retard African progress toward porportional representation in government, but they do not rule out the possibility of joint government and of sitting around a table with people of another race. Whether they see the sheer impracticability of it or the moral implications, there is no effort in Southern Rhodesia to set up Bantustans and to segregate the country in regional terms. Of course, certain lands are reserved for the whites and others for the Africans, but it is not suggested that the African lands should become self-sufficient units. Nor is the transfer of some white land to African use deemed impossible as it is in South Africa. It may be said that there is more flexibility in the Southern Rhodesian approach, and although the policies are by no means acceptable to the African majority, they are nevertheless more palatable than those of the Republic.

The character of the Federation as an intermediate between independent black Africa and South Africa is perhaps most clearly shown by the racial composition of the respective governments. In the Congo and Tanganyika, it is virtually all African. In South Africa, it is all white. In the Federation, the white minority still dominates the African majority, but Africans and whites sit together in the governing body. In the Congo and Tanganyika, African nationalism has attained its immediate goals. In South Africa, it may until now be said to have failed. In the Federation, it has achieved much and is making further progress.

The territories of Angola and Moçambique, though part of the buffer zone, perform their separating function in a different manner. It is hardly possible here to speak of transition between north and south; Portugal has created conditions which are not really comparable to those either to the north or south. While, however, there is no real transition as in the case of the Federation, there are important differences between these Portuguese areas and their northern and southern neighbors. From what has been said in a previous chapter, the contrasts between independent Africa and Portuguese rule are obvious. Yet, the Portuguese whites can and do accept an "assimilated" African virtually as an equal; much more so than the average Rhodesian, for instance. While the Portuguese do not harbor the racialist attitudes of so many other whites in Africa, they have nonetheless failed to give the African any significant power in the government of the African territories. What provides such tremendous contrast between Portuguese Africa and its neighbors has long been the condition of the "uncivilized" African in Overseas Portugal. Labor-conscription methods, wage levels, and literacy rates are among the many aspects in which Portuguese Africa compares unfavorably with its neighbors, north as well as south. Portugal's recent decision to eliminate the centuries-old distinction between civilized and uncivilized, assimilado and "unassimilated" Africans will not eliminate the effects of neglect and oppression. Whatever the privileges of the Portuguese voter may be, they are not likely to satisfy the Africans who are being made aware of what they have lacked for generations.

Portuguese Africa is experiencing violent rebellion against Lisbon's rule, and it is paying the price of being part of the buffer zone. While the Congo was in Belgian hands and Tanganyika under British rule, the events which began March 15, 1961, in northern Angola were not likely to occur. Within months after the Congo's independence, however, it became a haven for refugees from Angola. Leopoldville soon was the headquarters for the rebel Union of the Populations of Angola (UPA). With weapons readily available, the BaKongo people divided by the Congo-Angola border, and northern Angola forested and vulnerable to guerilla warfare, the outbreak of open rebellion was no surprise. The north of Moçambique, likewise, has experienced the effects of proximity to independent black Africa. And only Portugal's countinuing presence in Moçambique and Angola prevents similar friction from occurring on the borders of the Transvaal and South West Africa. Should Portuguese control over Moçambique and Angola fail, South Africa will no longer benefit from the protection of the buffer zone.

It is natural for the interested observer of the African scene to ask what the future is likely to bring. Predictions concerning the political evolution of Africa have rarely come true in the past, and making them is a hazardous business. There can be no doubt, however, that Southern Africa will see considerable political change, perhaps not only in terms of governmental representation and organization but even in boundary lines. In Nyasaland the possibility of joining a future East African Federation is being considered by African leaders who are determined to take their territory out of the Central African Federation. The eventual fragmentation of elongated Moçambique would appear likely as the buffer zone succumbs to the Wind of Change. South West Africa, with its Caprivi proruption, will undergo changes, and probably not only in political status. Its foremost port, Walvis Bay, remains a part of the Cape Province, hundreds of miles away. The northern border of this territory, where the Ovambo are divided between Angola and South West Africa and where the militant Herero are concentrated, may be the scene of future friction as the north Angola border has been since early in 1961. Major problems which may lead to far-reaching changes also confront the High Commission Territories, which will find themselves increasingly isolated as they develop toward independence. The nature of their political evolution has until today been diametrically opposed to that of mighty South Africa, while their dependence upon the giant neighbor has not been reduced. This must lead to difficulties, and it will require very skillful diplomatic action on the part of Britain to prevent the absorption of the High Commission Territories by South Africa.

One of the most commonly asked questions pertains to the Republic itself. For years now, observers have been speculating that a mass uprising must take place, and when the news of Sharpeville was first reported, many spoke of this as the beginning of the end for Nationalist rule. It is unlikely that such violent widespread uprisings as would bring an end to Afrikaner domination in South Africa will occur while the buffer zone survives. At present, South Africa's whites face only sporadic African opposition within the Republic itself, and from outside the country, little aid comes to that opposition. Africans in the Federation of the Rhodesias and Nyasaland are much too concerned over matters in their own country to be concerned over South Africa. The Portuguese have controlled their territories very tightly, and if any assistance is the form of organizers and arms have percolated to South Africa, the quantity has been very small. Within the Republic, the whites have clamped severe restrictions upon the Africans; they have outlawed organizations which might develop into dangerous cores of opposition, and they have placed the automatic death penalty upon any African found possessing firearms. Thus it may be said that South Africa's control over the African majority remains effective. When, however, independent African states have developed adjacent to South Africa itself, when the Wind of Change reaches South Africa's very borders, an altogether different situation will exist. At that time, South Africa will face not only the opposition of Africans within its own borders but also the full onslaught of all Africa to the north. It is, therefore, of vital importance to South Africa that the Portuguese and British retain some form of control in Overseas Portugal and the Central African Federation. This explains the voting in several important United Nations assemblies, and it also manifests itself in Portuguese-South African military cooperation, which has been much strenghtened over the period since late 1960. The close trade connections between South Africa and Southern Rhodesia and the fact that many whites living in Southern Rhodesia have come from South Africa indicate why there is a core of Afrikaner sentiment in the Federation. It is in the interest of the whites to maintain the status quo.

White South Africans frequently show an awareness of the difficulties they will face when African independence comes to knock on their doors. They point out that while Africa has been in a turmoil, South Africa has been relatively quiet, and that Africans in the Republic have shown their satisfaction with their lot by their failure to rise in really great numbers. One is reminded that, even during the weeks following the shooting at Sharpeville, when general strikes were called, most Africans did go or wanted to go to work and that most of those who stayed home were intimidated by threats. Indeed, any objective observer of the African scene must conclude that in South Africa a really widespread, general uprising on the part of the Africans is a most unlikely event while the present political pattern remains. The South African interpretation of this fact is that the Africans are generally satisfied with what they have, that they do not object seriously to segregation, and that while some may occasionally object to such practices as the pass laws, most are aware of the necessity of these things and accept them thus. Hence, until these people are stirred up, not by the conditions under which they live but by outside agitation, they will remain law-abiding citizens of the Republic. This interpretation is ground for the bitterness with which South Africans resent the failure of some colonial powers and the United States to render support in the United Nations and elsewhere. The facts, they argue, clearly show that in an Africa of chaos, a stable state has been created. In some independent African countries Africans have treated other Africans with no more circumspection than that with which white South Africans are treating Africans. Nkrumah's activities with reference to the Ghana Opposition made big headlines in South Africa. Was this the kind of democracy for which whites had to abandon their rule? Did whites, in order to gain the favors of these new African oppressors, have to fail in their support of people of their own race, religion, and culture? What reason was there to favor African nationalism over white nationalism? What evidence did the Congo give of Africans being better off after the colonists had gone? Africans were starving in Kasai Province—where in White-ruled Africa did Africans starve to death in 1961? Where in all Africa did Africans have the housing, health facilities, and educational opportunities they have been given in the Republic? What grounds, therefore, are there to abandon all that has been achieved in favor of the near-totalitarian rule of another Nkrumah, Touré, or Lumumba?

Indeed, conversations with some of the Africans living in the better housing developments of Johannesburg's African suburbs will provide the answers South Africa's whites like to hear. Some failed to join the strikes of Black Week in 1960 because they did not want to lose their jobs, relatively high pay, and such marvels as a tap right in their own house, meat supplies at the corner market, a radio in the living room, and the like. "You cannot eat that vote they are fighting for," several commented. These are the statements the whites take as being evidence of the general mood among the great majority of the African population. The utter failure of the general strike called just prior to the establishment of the new Republic was taken as further evidence that opposition to South Africa's government is not a popular feeling among the country's 10 million Africans.

It is a tragedy that South Africa's white leaders are permitting themselves to be misguided by the acceptance of a few of these statements and the willing cooperation of some lackey chiefs. The truth is that there is no real measure of the mood of South Africa's Africans, because their avenues of expression have been closed. They are not consulted on matters vitally affecting the whole country, have no vote, and are controlled by pass laws and firearms. Some have benefited from South Africa's general prosperity, and many of those who have do not wish to lose their favored position by involving themselves in political activities. They are used as propaganda puppets, but in terms of prevalent opinion they are by no means the representative group South Africans like to believe them to be. Some whites have permitted themselves to be deceived to such an extent that they really believe that agitation from across the borders will not find a response among South Africa's Africans.

It is not likely that Africans either within or outside of South Africa will rest before the Republic has been made to change its ways—and that is probably the mildest possible statement of the Africans' goals. Even if South Africa were the paradise some Afrikaners believe it to be in comparison with the rest of the continent, it is not likely that the Africans would have been satisfied. It is no longer enough to point to the big schools, large hospitals, rising wages, and improved living conditions to justify the political situation of the non-whites. Informed Africans are fully aware that several African countries have seen a setback in such conditions as well as their economy after independence, but, they argue, Africans are at the helm, and that is all that matters. South Africa, therefore, will have to do something beyond carrying out its much-advertised scheme of parallel but separate development. It will have to convince Africans in the remainder of Africa of its own propaganda—that the plan really does aim at equality in addition to separation and that Africans in South Africa do want to participate in its implementation. In this, South Africa is not likely to succeed, and as long as this is so, the country will face the wrath of many millions, and not in Africa alone. For this, Afrikaners have largely themselves to blame. Apartheid has been set forth as a doctrine in the most uncompromising terms. Many moves in its execution have been accompanied by hotheaded speechmaking in a tradition which has helped the Nationalists attain a worldwide reputation for undiluted racist extremism. Long before Apartheid, South Africa was well on its way toward a form of segregation, one which was no less objectionable to the Africans but one which at least was capable of a melioration with time, since it was not everywhere rigidly imposed by law. Now, everything has been enforced by law, and in every walk of life South Africans of all races are each day involved in it, unable even if they privately desire to circumvent it. Someone in the government decided in 1960 that there should be no handshakes between black and white, and next day the newspapers were full of pictures: an African was to be greeted by holding up one arm from the elbow, but no contact was to be made. Soon, someone else repealed the idea, but meanwhile one may contemplate what was the effect upon educated Africans particularly, of finding this new rule so crudely described in the white man's papers. This sort of thoughtlessness is going to have its effects when South Africa's hour of crisis arrives, and the needlessness of it all adds immeasurably to the tragic aspect of that inevitable circumstance.

Reflecting upon the unequaled hospitality of the Afrikaners, the jovial sportsmanship of English-speaking

South Africans, the patience of the Africans, the diligence of the Asiatic South Africans, and the cheerfulness, under stress, of the Colored people, it is difficult to understand just how human beings with such great assets have descended into an abyss of such hopeless division. What happens in the buffer zone remains to be seen, but South Africa appears to be beyond repair, and when one looks back to see what chances have been missed here, one is filled with dismay. This beloved country, as Paton has aptly called it, has so much more than a normal share of the world's riches —gold, diamonds, good soils, and many more—and yet has so much more than a reasonable share of unhappiness. It is not overpopulated, yet because of current policies parts of it may become so. It is economically sound but faces a crippling boycott. It could have been a showcase of interracial cooperation and has become a hotbed of racism. It could have retained a position of leadership in the Africa of tomorrow but now faces a clouded future. For this, it would be unfair for only one section of South Africa's multiracial population to be held responsible, nor is the present generation of South Africans alone at fault. South Africa's problems have deep, age-old roots. The only remedy that might have solved them is time, many decades of it. In the fast-changing Africa of today, that commodity is no longer available in quantity. Britain, France, Belgium, and Portugal have discovered this and display a varying degree of flexibility in dealing with the new condition. Flexibility is not one of the characteristics of South Africa. The consequences must be clear.

Index